Christian Schlieder

Autodesk® Inventor® 2011

Basics in theory and practice

Many practical exercises on the
construction object four-stroke-engine

Christian Schlieder

Autodesk® Inventor® 2011

Basics in theory and practice

Many practical exercises on the construction object four-stroke-engine

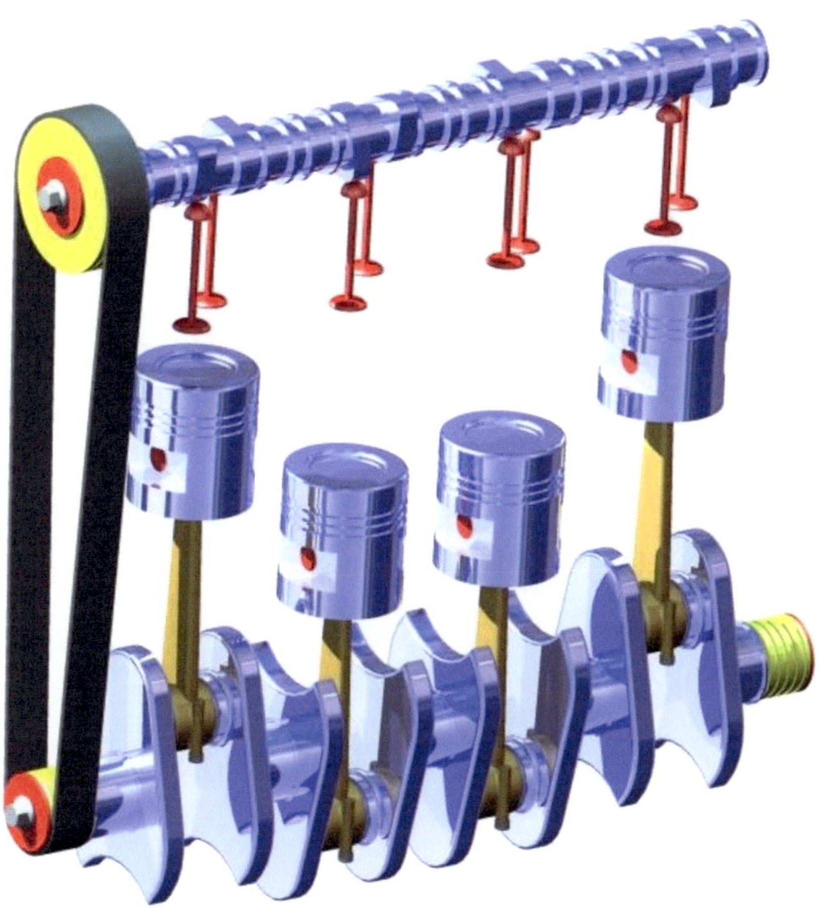

Autodesk
Authorised Author

Dipl.-Ing. Christian Schlieder is authorized Autodesk[®] Author. He is a teacher for various CAD- Programs in Germany.

The information contained in this book has been checked in all conscience. As mistakes cannot be excluded neither author nor publisher accept any responsibility, liabilities or warranties of any kind nor liability for the use of the provided information. The book simply shows an excerpt of the program *Autodesk*[®] *Inventor*[®] *2011*. The entire program is not dealt with in this book (the working of this book does not lead to expertise in working the entire program). The used example engine is completely fictional and not functional. The book is based on a German basis and was translated into English. The sample engine originally was drawn in metric system (kilograms, millimeters, Newton's, seconds). However, you can use any other system. To avoid confusion during the editing phase of this book some of the measures have been skipped. Simply type the values into the program areas (without measures). Author and publisher do not warrant that the described procedures or processes are free from Third Party rights. The book is protected by intellectual property rights. Translation, reprint, copy and other use of the book or parts thereof are not permitted without written permission by the author. *Autodesk*[®] *Inventor*[®] *2011* is a registered trademark of Autodesk, Inc., and/ or their affiliates and/ or their subsidiaries in the USA and other countries.

ISBN

9783842345621

IMPRINT

Dipl.- Ing. Christian Schlieder
www.Ingenieurbuero-Schlieder.de
Fax: +49 (0) 3212 - 1122290

PRODUCTION AND PUBLISHER

Books on Demand GmbH, Norderstedt
www.BoD.de

CONTENTS

1   Handling the book

1.1   Target group & scheme of the book

This book is directed to all interested people of all areas of studies. It is set up logically and tries to bring the program **Autodesk® Inventor® 2011** by the successive construction of a **four-stroke-engine**. In small snippets the reader will get to know different procedures and commands and rework in steps. You hold an exercise and fundamental book for **Autodesk® Inventor® 2011** in your hands. Some parts are explained in detail others only find short mention. It is advised to sporadically leave the chain of commands and intuitively start your own tests. The program helps you with this with plenty of help and instructions.

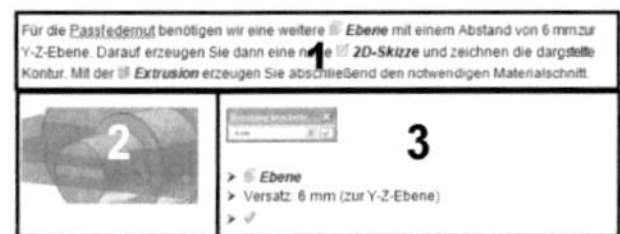

Every segment includes a text explanation (1) followed by a graphic depiction of the procedure (2) and a short chain of commands (3). Please note the references!

It was attempted to use as little text as possible and plenty of graphic depictions. Much of it is self-explanatory by trying. Be courageous and try various commands. Following abbreviations are used throughout the book to simplify the chain of commands:

➢ **D**	Diameter
➢ **R**	Radius
➢ **ESC**	Keyboard-key **Escape**
➢ **Strg+C**	Keyboard-keys **Strg** and **C**
➢ **Strg+V**	Keyboard-keys **Strg** and **V**
➢ **LMC**	Left mouse click
➢ **RMC**	Right mouse click

1.2   Digital accessories of the book

All created components, building blocks and drawings are available at following website at no cost:

➢ **www.Ingenieurbuero-Schlieder.de/html/Download-EN.html**

Use the data to get ideas for the construction or to load components not mentioned in the book.

2 Program structure & program management

2.1 In regards to program structure

Program operation of **Autodesk® Inventor® 2011** has been well adjusted to the user. Symbols and commands are already split into logical clusters (ribbons). You will find all respective commands easily and quickly. Also many introductions, tutorials, demo-animations and a very good handbook are included. With the available help the user is capable to learn **Autodesk® Inventor® 2011** auto didactical. Should you have been using Autodesk® Inventor® before the 2010 version you surely are used to the known classical view which can be set under following chain of commands:

> ➢ **Tools > Application Options > Colors > Interface Style: Classic UI**

Tip: Please note the ⑦ **Question mark (Help)**, which is located in the lower left part of the corresponding window of many commands. Here you also are directed to the respective explanation in the **Autodesk® Help**. Alternatively you can also use the **F1**-key in many cases.

2.2 User interface

Amongst others the basic commands like **Save**, **Open**, **Print** und **Close** and **Export** (e.g. exporting data files into other formats), **Manage** (e.g. projects, iFeature-catalog, construction assistant), **iProperties** (all building block properties) and **Server** (Register: Safe) are included in the main menu.

2.2.1 The main menu

Following commands you will find in the main menu:

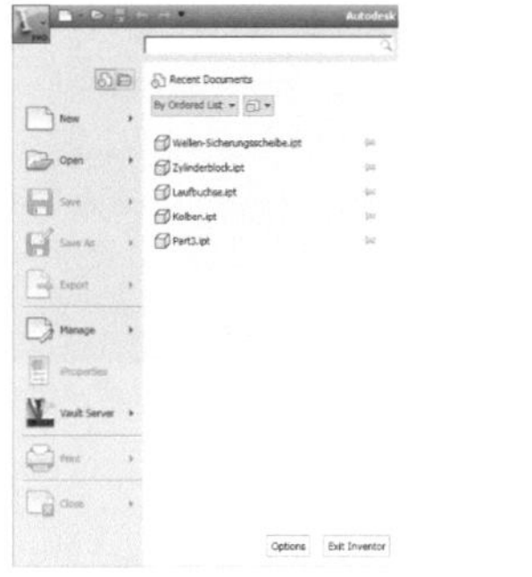

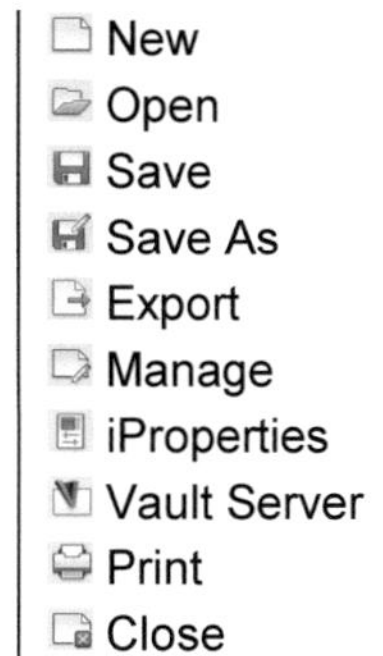

New
Open
Save
Save As
Export
Manage
iProperties
Vault Server
Print
Close

2.2.2 The multifunction bar

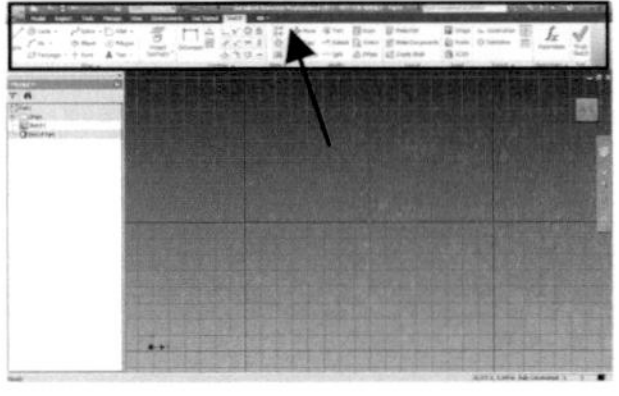

On the **multifunction bar** you find logically ordered command clusters which are assigned to the corresponding register cards. You have the option to blend and hide the clusters/groups in the registers. Simply right click with **RMC** on the **group** and place the corresponding hook.

2.2.3 The Browser window

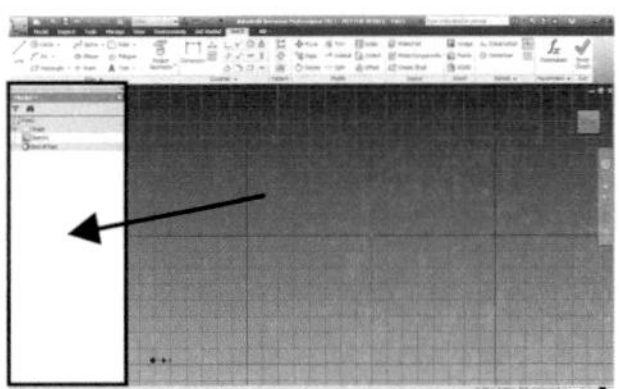

In the **Browser window** the process history of the respective file is listed. Here you have the option to change or delete the processes steps executed in the file.

2.2.4 The graphic window

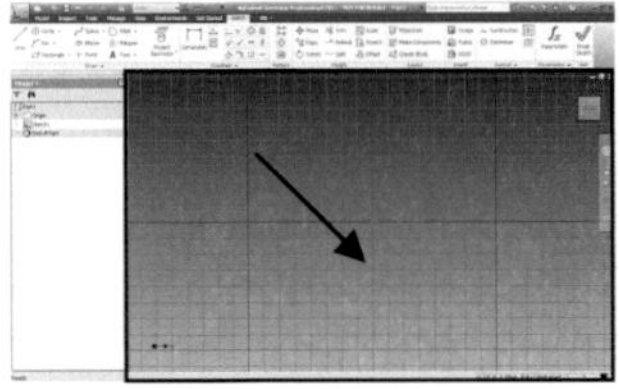

The **graphic window** contains the actual status of the created object. This is the main work area. The graphic window opens automatically when you open a new or existing file.

2.2.5 View Cube, Steering Wheels & the navigation bar

The **View Cube** is a steering element to change the respective view of the object in the graphic window. You can choose the view by clicking on the corresponding button or freely turn it in the room by pressing with **LMC** on the View Cube, keep the key pressed and then move the mouse.

Tip: An alternative to the freely turning of the objects in the graphic window is the combination of the ⇧ Shift shift-key with the middle mouse key while moving the mouse.

Steering Wheels (complete navigation wheels) combine many of the available navigational tools on a surface.

Navigation bar contains various indication and navigation commands. Here you will also find ***View Cube*** and ***Steering Wheels*** amongst others. A short overview of the commands:

Full Navigation Wheel	Previous View
Pan	Next View
Zoom All	Perspective
Free Orbit	Orthographic
View Face	Shaded
Zoom Window	Ground Plane
Zoom Selected	

2.2.6 General commands of the home page

In this toolbox you will find the commands for ***Search*** and ***Help.*** With the Autodesk® help system you will find solutions to problems or questions quick and uncomplicated, simply type in a key word or a question. Here a short overview over the commands:

New	Select Face and Edges
New Assembly	Select Sketch Features
New Drawing	Select Wires
New Part	Color Override
New Presentation	Keyword or Phrase
New *.dwg	Search
Open	Subscription Center
Save	Communication Center
Undo	Favorites
Redo	Help
Update (Local)	Minimize
Select Bodies	Restore Down
Select Groups	Close
Select Features	

2.3 Register FIRST STEPS

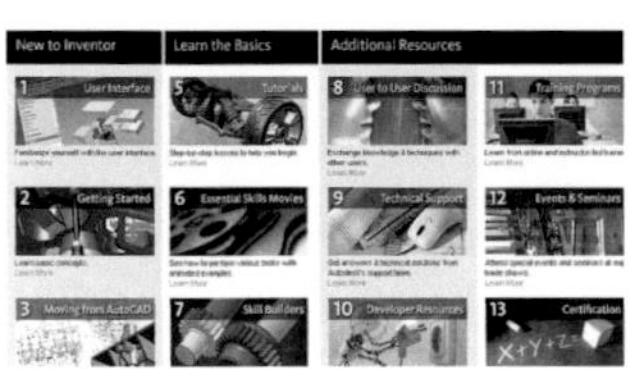

Here you find valuable help like, e.g. learning programs, tutorial videos for the new *Autodesk® Inventor® 2011* or the new multifunction bars, overviews over new functions, a handbook, Demo-animations and much more.

2.3.1 Command overview Register FIRST STEPS

New Open Projects Ribbon Introduction Ribbon Tutorial Command Locator What's New Getting Started Tutorials Learning Resources Show Me Animations Enginees Rule.ORG Wiki Help Customer Involvement Finish Sketch

| Launch | User Interface Overview | New Features | Learn about Inventor | Community | Exit |

New
Open
Projects
Ribbon Introduction
Ribbon Tutorial
Command Locator
What´s New
Getting Started

Tutorials
Learning Resources
Show Me Animations
Enginees Rule.ORG
Wiki Help
Customer Involvement
Finish Sketch

2.4 Register TOOLS

The register *Tools* is used for modification of the user and document settings. The commands *Streamline* and *Supplier content* offer an online-service for project management and cooperation between companies and their partners, customers or suppliers.

2.4.1 Command overview Register TOOLS

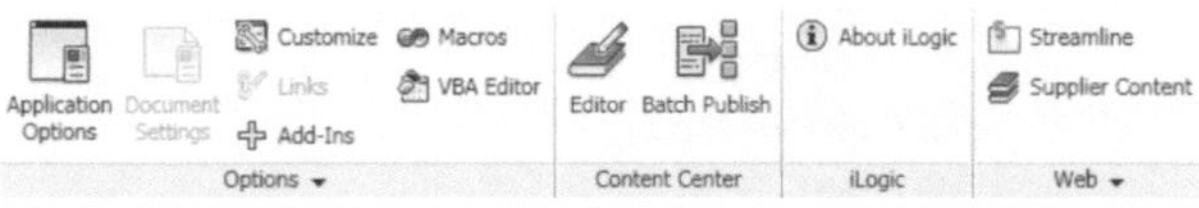

▣ Application Options	▣ VBA Editor
▣ Document Settings	▣ Editor
▣ Customize	▣ Batch Publish
▣ Links	▣ About iLogic
✛ Add-Ins	▣ Streamline
▣ Macros	▣ Supplier Content

Tip: The command ▣ **Update** is located in the upper program bar and usually has a grey background. It is only activated when there is something to refresh in the graphic window. Pay attention to the activated symbol.

---

3 PROJECT MANAGEMENT & BACKUP

3.1 MANAGING PROJECTS

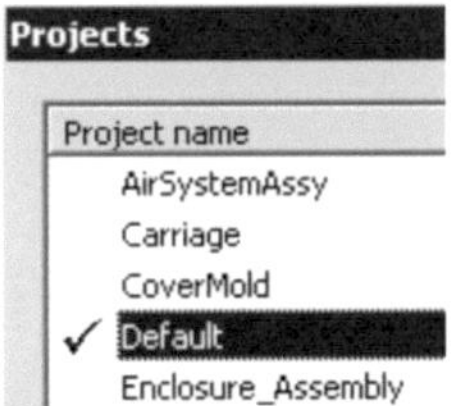

Working with **Autodesk**® **Inventor**® **2011** should principally be done in **projects**. A project creates a work area in which project related data (files, backup copies, project information) are saved. Individual people or groups of people can work on one project.

3.2 BACKUP

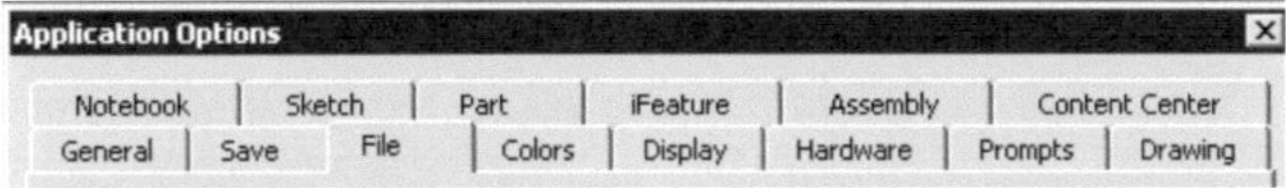

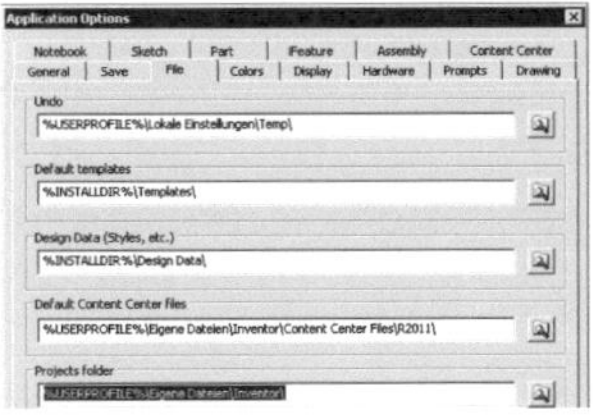

It is important to check the current storage location of the project file and to have a correct project management. Under:

> Register: **Tools**
> **Application Options** > **File** > **Projects Folder**

You can define the already chosen project folder in project creation. Therefore also future automatic saving of the program in the proper folder is ensured. The program automatically creates security backups of your files (folder: **Old Versions**). Should you have forgotten to save you can search for current versions of your files here.

4 BASICS & GENERAL COMMANDS

4.1 Creating a new FILE

➢ Register: **Get Started > New**

In this area of command you have the option to choose from the following Standard-files:

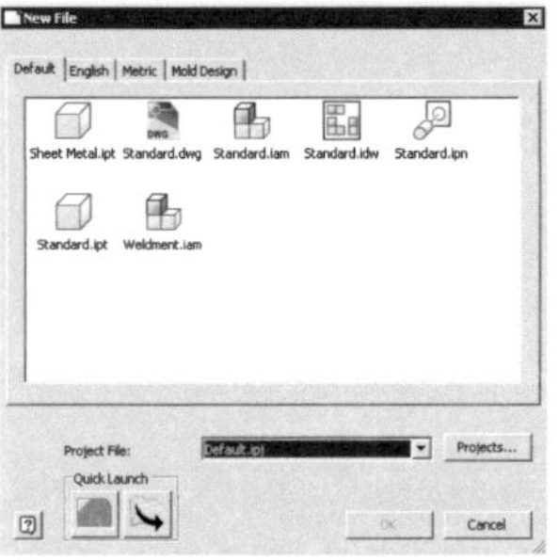

➢ **SheetMetal.ipt** creates a sheet metal part
➢ **Standard.dwg** creates and Autodesk® AutoCAD®-drawing (.dwg Format)
➢ **Standard.iam** creates an assembly
➢ **Standard.idw** creates an Autodesk® Inventor®-drawing (.idw Format)
➢ **Standard.ipn** creates a presentation
➢ **Standard.ipt** creates a part
➢ **Weldment.iam** creates an assembly as a weld construction

The program creates a file from the chosen template. You also can choose an already existing project.

Tip: You can create your own templates and save them.

4.2 SKETCHES

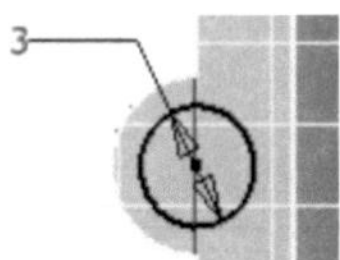

Sketches constitute the construction basis for almost all construction elements. Here the basic outlines are created which later are transformed into volume frame with 3D-commands in the area **Parts.**

4.2.1 Command overview Register SKETCH

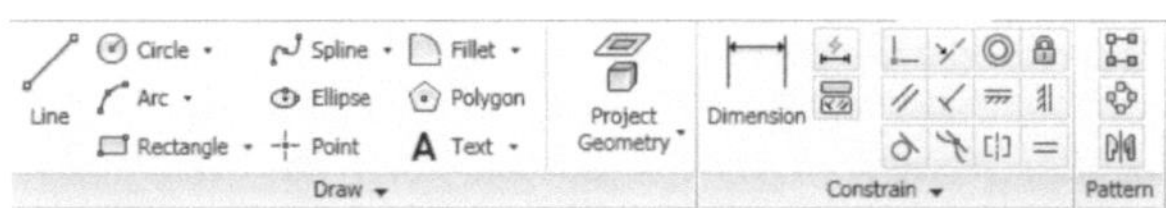

/ Line
⊘ Circle (Center Point)
◯ Circle (Tangent)

⤴ Automatic Dimensions and Constraints
▦ Show Constraints
▦ Edit Coordinate System

- Arc (Three Point)
- Arc (Tangent)
- Arc (Center Point)
- Rectangle (Two Point)
- Rectangle (Three Point)
- Spline
- Bridge Curve
- Ellipse
- Point
- Fillet
- Chamfer
- Polygon
- **A** Text
- Geometry Text
- Project Geometry
- Project Cut Edges
- Project Flat Pattern
- Dimensions

- Constraint Inference
- Constraint Persistence
- Constraint Dimensions
- Coincident Constraint
- Collinear Constraint
- Concentric Constraint
- Fix
- Parallel Constraint
- Perpendicular Constraint
- Horizontal Constraint
- Vertical Constraint
- Tangent
- Smooth (G2)
- Symmetric
- Equal
- Rectangular Pattern
- Circular Pattern
- Mirror

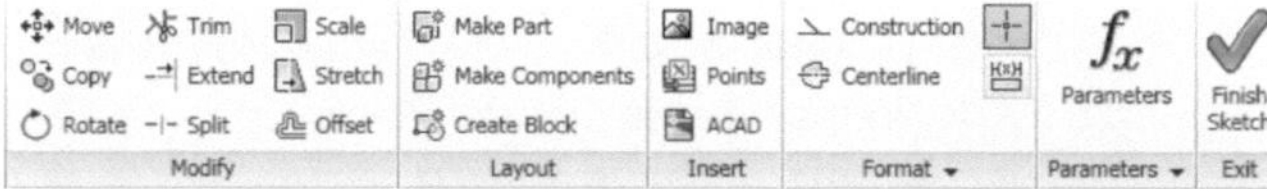

- Move
- Trim
- Scale
- Copy
- Extend
- Stretch
- Rotate
- Split
- Offset
- Make Part
- Make Components

- Create Block
- Image
- Points
- ACAD
- Construction
- Point
- Centerline
- Driven Dimensions
- Parameters
- Finish Sketch

4.3   PARTS

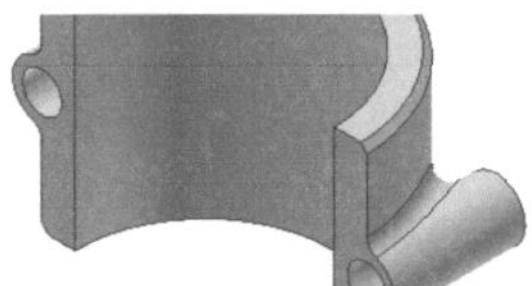

In the area **Parts** sketches are transformed into new volume frames (components), existing components are edited or further features are used.

4.3.1   Command overview Register MODEL

- 2D Sketch
- 3D Sketch
- Extrude
- Revolve
- Loft
- Sweep
- Rib
- Coil
- Emboss
- Derive
- Hole
- Fillet
- Chamfer
- Thread
- Move Face
- Shell
- Split
- Copy Object
- Draft
- Combine
- Move Bodies
- Bend Part
- Plane
- Offset from Plane
- Parallel to Plane through Point
- Midplane between Two Parallel Planes

- Midplane of Torus
- Angle to Plane around Edge
- Three Points
- Two Coplanar Edges
- Tangent to Surface through Edge
- Tangent to Surface through Point
- Tangent to Surface and Parallel to Plane
- Normal to Axis through Point
- Normal to Curve at Point
- Axis
- On Line or Edge
- Parallel to Line through Point
- Through Two Points
- Intersection of Two Planes
- Normal to Plane through Point
- Through Center of Circular/ Elliptical Edge
- Through Revolved Face or Feature
- Point
- Grounded Point
- On Vertex, Sketch point, or Midpoint
- Intersection of Three Planes
- Intersection of Two Lines
- Intersection of Plane/ Surface and Line
- Center Point of Loop of Edges
- Center Point of Torus
- User Coordinate System

Rectangular Pattern

Circular Pattern

Mirror

Thicken/ Offset

Boundary Patch

Stitch Surface

Trim Surface

Sculpt

Delete Face

Extend

Replace Face

Copy to Construction

f_x Parameters

Import from XML

Export to XML

Grill

Snap Fit

Boss

Rule Fillet

Rest

Lip

Distance

Angle

Loop

Area

Region Properties

Derive

Decal

Place Feature

Insert iFeature

Insert Object

Place iFeature from the iFeature Catalog

Import

Place Pin

Place Pin Group

Harness Properties

Create iPart

Edit using Spreadsheet

Edit Factory Scope

Stress Analysis

Convert to Sheet Metal

4.4   ASSEMBLIES

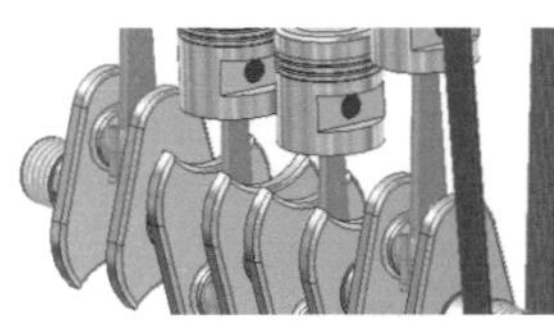

Here **assemblies** are created. You can insert existing parts, create new parts, generate constraints and use further features.

4.4.1   Commands overview Register ASSEMBLE

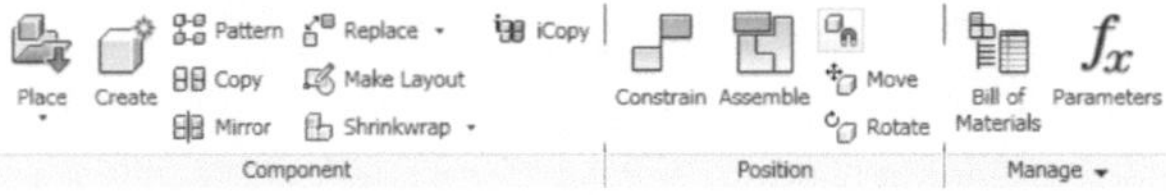

Place
Place from Content Center
Place iLogic Component
Create
Pattern
Replace Component
Replace All Components
Copy
Make Layout
Mirror
Shrinkwrap

Shrinkwrap Substitute
iCopy
Constrain
Assemble
Grip Snap
Move Component
Rotate Component
Bill of Materials
f_x Parameters
Import from XML
Export to XML

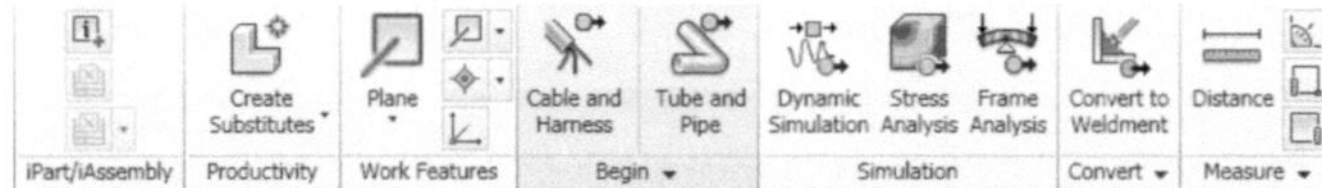

iPart/ iAssembly Author
Edit iAssemblies
Edit using Spreadsheet
Edit Factory Scope
Create Substitudes
Update Substitudes
Save and Replace Component
Add Part
Add Assembly
Link Levels of Detail
Component Derive
Place at Component Origin
Rename Browser Nodes
Alpha Sort Component
Ground and Root Component
Degree of Freedom Analysis
Cable and Harness
Tube and Pipe

ISOGEN Output
Define Gravity
Tube and Pipe Styles
Dynamic Simulation
Stress Analysis
Frame Analysis
Convert to Weldment
Fillet Weld Calculator (Plane)
Fillet Weld Calculator (Spatial)
Plug/ Groove Weld Calculator
Butt Weld Calculator
Spot Weld Calculator
Bevel Solder Calculator
Butt Solder Calculator
Lap Solder Calculator
Step Solder Calculator
Step Tube Solder Calculator
Bead Report

4.4.2 Commands overview Register DESIGN

Register **Design** contains the command clusters: **Fasten** (screw connection, bolts), **Frame** (Insert, Change, Miter, Notch, Trim/ Extend, Lengthen/ Shorten, Analysis), **Power Transmission** (Shaft, Spur Gear, Bearing, V- Belts, Key, Disc Cam, Parallel Splines, O-Ring), **Spring** (Compression, Extension, Belleville, Torsion) and **Measure**.

4.4.3 Command overview Register INSPECT

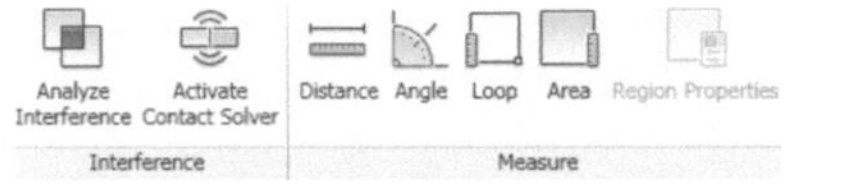

Register **Inspect** *contains the command clusters*: **Interference** (Analyze Interference, Activate Contact Solver) and **Measure**.

4.4.4 Command overview Register TOOLS

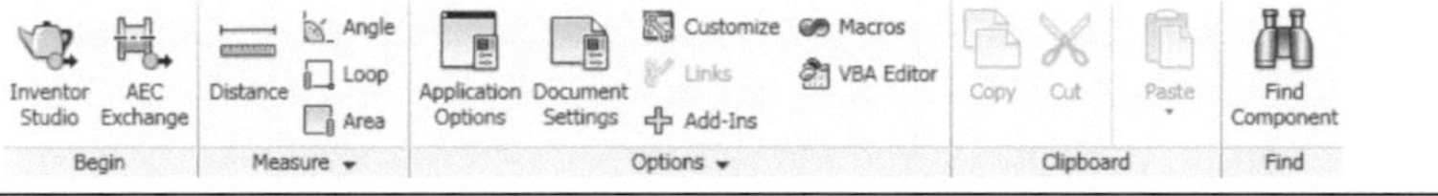

Register **Tools** contains the command clusters: **Begin** (creation of high resolution graphics), **Measure**, **Options** (changing of individual program settings), **Clipboard** and **Find** (Search for a component in the Autodesk® Contents center).

4.4.5 Command overview Register MANAGE

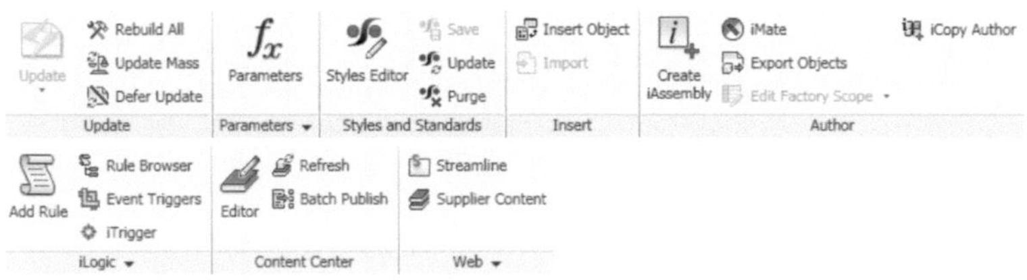

Register **Manage** contains the command clusters: **Update** (Refreshing of existing components), **Parameters**, Styles and Standards, Insert, **Author**, **iLogic**, Content Center (editing of existing or creation of new content center- library parts) and **Web** (Streamline and Supplier Content for project management and cooperation).

4.4.6 Command overview Register VIEW

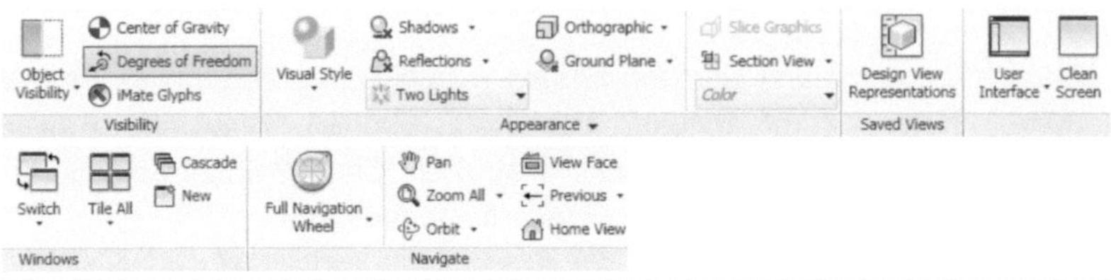

Register **View** contains the comand clusters: **Visibility** (Object Visibility, Center of Gravity, Degrees of Freedom, iMate Glyphs), **Appearance** (Visual Style, Shadows, Reflections ...), **Saved Views**, **Windows** and **Navigate** (Navigation Wheel, Zoom, Orbit ...).

4.4.7 Command overview Register ENVIRONMENTS

Register **Environments** contains the command clusters: **Begin** (Dynamic Simulation, Stress- and Frame- Analysis, Inventor Studio, AEC Exchange, Cable and Harness, Tube and Pipe ...), **Convert** (Convert to Weldment) and **Manage**.

4.5 DRAWINGS

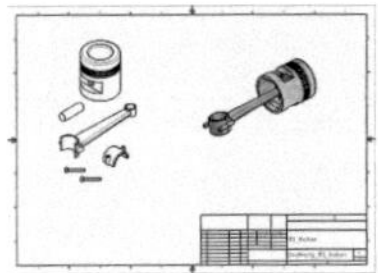

In the area of **Drawings** parts or assemblies previously created can be derived as drawings.

These can later be exported in various formats.

4.5.1 iProperties, work sheets & drawing resources
4.5.1.1 iProperties

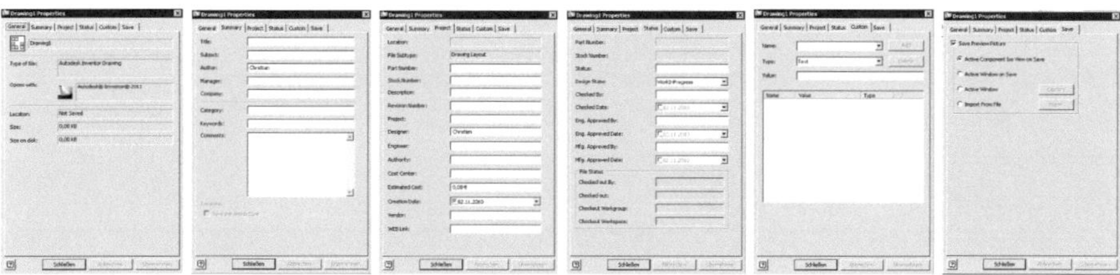

4.5.1.2 Drawings

Drawings contain standard frames and standard text fields. Both can be edited (**Drawing:1** > **RMC** > **Edit Sheet**) on a respective object or adding new templates.

4.5.1.3 Drawing Resources

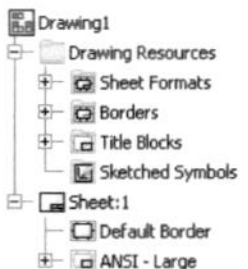

Drawing Resources contain standard collections of text fields, frames, symbols, types of views and further standards. Next to the already existing values you can create individual templates or edit the existing ones.

4.5.2 Command overview Register PLACE VIEWS

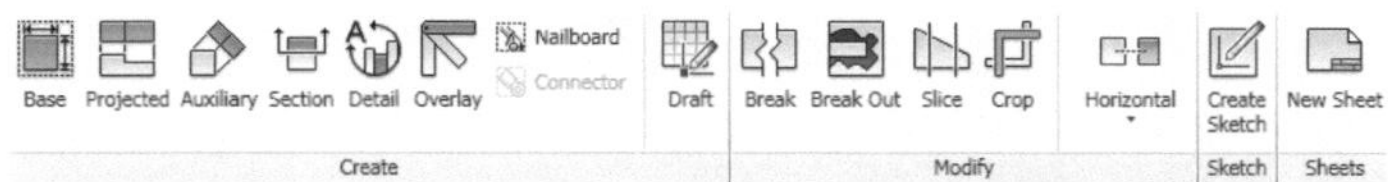

Base	Break Out
Projected	Slice
Auxiliary	Crop
Section	Horizontal
Detail	Vertical
Overlay	In Position
Nailboard	Break Alignment
Connector	Create Sketch
Draft	New Sheet
Break	

4.5.3 Command overview Register ANNOTATE

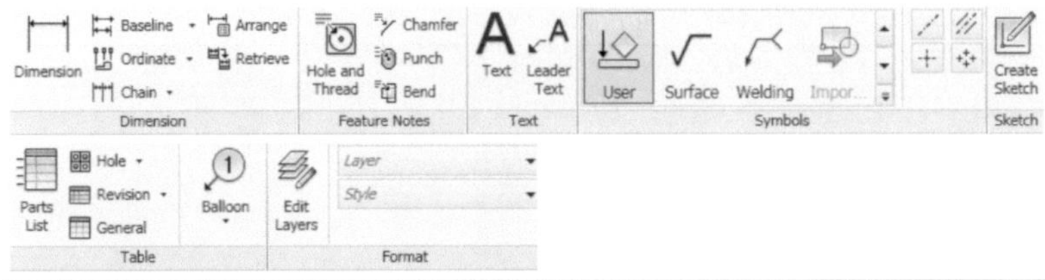

⊓ Dimension	√ Surface
⊢ Baseline	⊡ Import AutoCAD Block
⊢ Baseline Set	⟋ Centerline
⬚ Retrieve	⫽ Centerline Bisector
⊔ Ordinate	⊹ Center Mark
⊔ Ordinate Set	⊹ Centered Pattern
⬚ Arrange	⬚ Create Sketch
⊞ Chain	⊞ Parts List
⊡ Hole and Thread	⊞ Hole
⟋ Chamfer	⊞ Revision
⊚ Punch	⊞ General
⬚ Bend	① Balloon
A Text	⊕ Auto Balloon
⟋A Leader Text	⊛ Edit Layers
⬚ User	Select Layer/ Style
⟋ Welding	

4.6 PRESENTATIONS & INVENTOR STUDIO

4.6.1 Command overview Area PRESENTATIONS

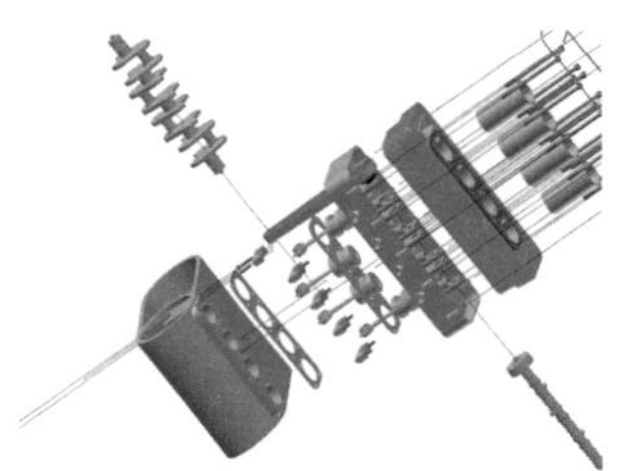

In the area **Presentations** you can create and animate exploded views.

Save your animations later as video-files or render a high-resolutions photo to conclude your presentation.

- **Create View**
- **Tweak Components**
- **Precise View Rotation**
- **Animate**

4.6.2 Command overview area *INVENTOR STUDIO*

With the command **Inventor Studio** (**Tools > Inventor Studio**) high-resolution photos are created, lighting and surface effects are edited or animations created.

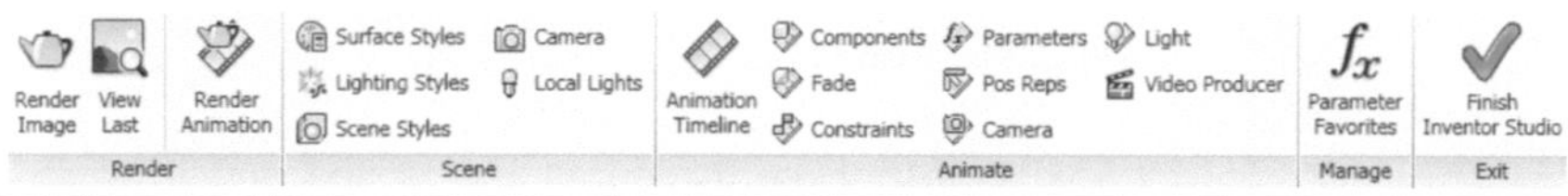

- Render Image
- View Last
- Render Animation
- Surface Styles
- Lighting Styles
- Scene Styles
- Camera
- Local Lights
- Animation Timeline
- Components
- Fade
- Constraints
- Parameters
- Pos Reps
- Camera
- Light
- Video Producer
- Parameters Favorites
- Finish Inventor Studio

4.6.3 Surface & Lighting styles

Surface Styles: Here you can create new surface styles, edit existing ones, import and export styles.

Lighting Styles: Edit the lighting options to properly light the components.

4.6.4 Cameras & component specific lighting

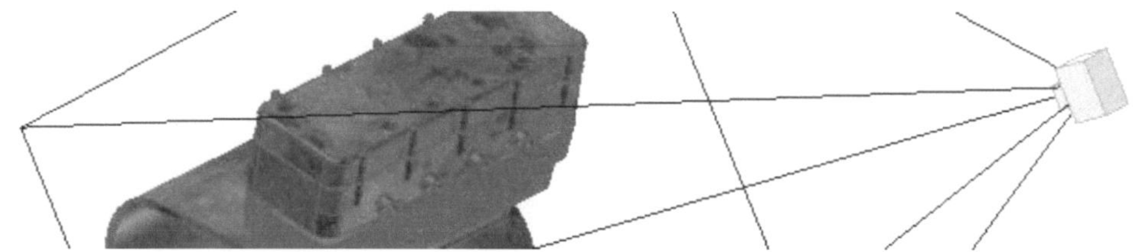

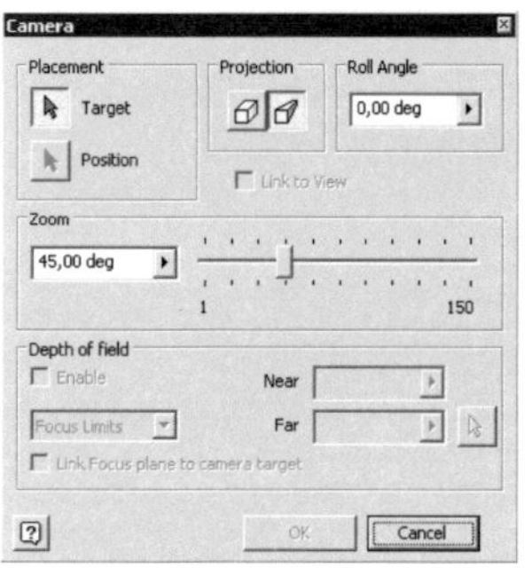

Camera: You can create new cameras to get an exactly defined view for example the ☞ **Render** on your components. Here you have the option to adjust various parameters (e.g. placing, zooming, rolling angle, near, far and glare values).

Local Lights: Here lights can be defined and specially aligned (e.g. as hotspots).

4.7 SHEET METAL AREA & WELDING CONSTRUCTIONS
4.7.1 Command overview area SHAPING OF SHEET

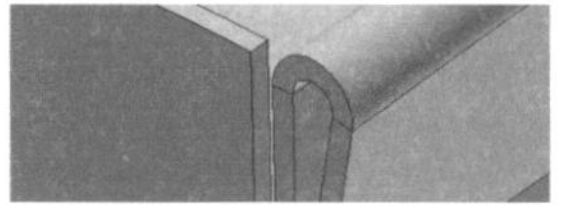

In the area **Sheet Metal** you can create and edit sheets and create **Unfolds**. There are useful tools like **Punch Tool**, **Hem** or **Flange**.

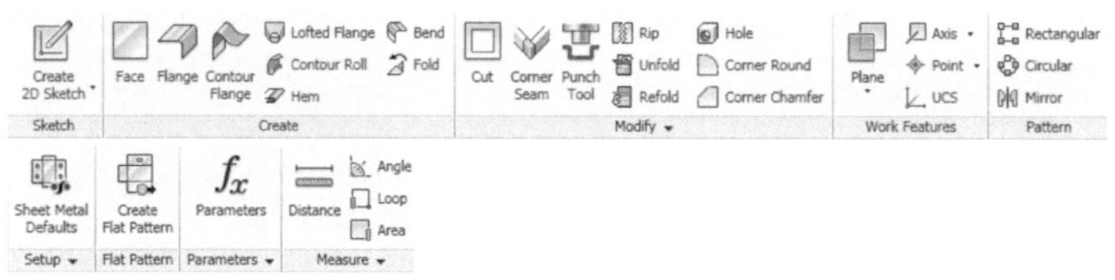

⬚ 2D Sketch	⬚ Refold
⬚ Face	⬚ Hole
⬚ Flange	⬚ Corner Round
⬚ Contour Flange	⬚ Corner Chamfer
⬚ Lofted Flange	⬚ Plane
⬚ Contour Roll	⬚ Axis

✐ Hem	✦ Point
☝ Bend	⌐ UCS
✌ Fold	⊞ Rectangular Pattern
☐ Cut	✪ Circular Pattern
✎ Corner Seam	◖◗ Mirror
⚒ Punch Tool	▣ Sheet Metal Defaults
▨ Rip	▤ Create Flat Pattern
▤ Unfold	f_x Parameters

4.7.2 Command overview area WELDING CONSTRUCTIONS

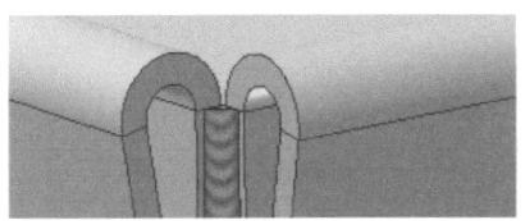

In the area **Welding Constructions** weld mends, weld mend reports and weld mend adaptions can be created.

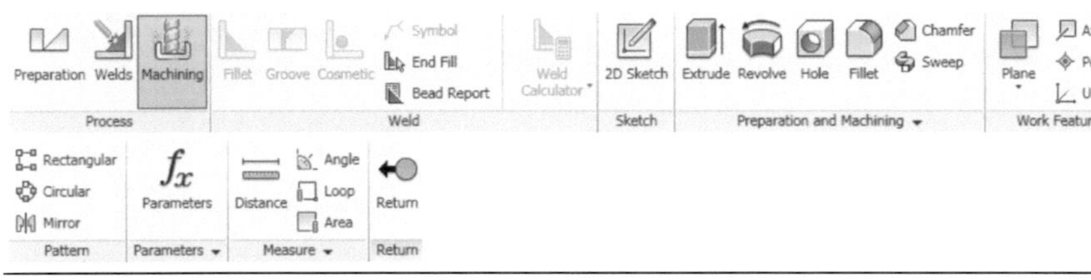

▱ Preparation	⊙ Hole
⬀ Welds	◗ Fillet
▥ Machining	◍ Chamfer
◣ Fillet Weld	✆ Sweep
⊓ Groove Weld	▱ Plane
▙ Cosmetic Weld	▨ Axis
⤹ Symbol	✦ Point
▥ End Fill	⊞ Rectangular Pattern
▤ Bead Report	✪ Circular Pattern
▚ Weld Calculator	◖◗ Mirror
▨ 2D Sketch	f_x Parameters
▥ Extrude	☋ Return
◎ Revolve	

4.8 STRESS ANALYSIS & PARAMETERS

4.8.1 Command overview area STRESS ANALYSIS

The command **Stress Analysis** is located in the register **Environments**. Here parts and assemblies can be tested for stresses. The results can later be exported and used further.

Create Simulation	Simulate
Parametric Table	Animate
Assign	Probe
Fixed	Convergence
Pin	Same Scale
Frictionless	Color Bar
Force	Probe Labels
Pressure	Maximum Value
Bearing	Minimum Value
Moment	Boundary Conditions
Gravity	Smooth Shading
Automatic Contacts	Adjust Displacement Display
Manual Contacts	Report
Mesh View	Guide
Mesh Settings	Stress Analysis Settings
Local Mesh Control	Finish Stress Analysis
Convergence Settings	

4.8.2 Area PARAMETERS

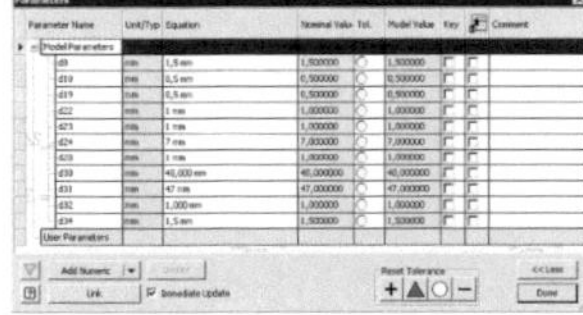

Parameters control component properties and place constraints among themselves.

With this you can control sketches, parts and assemblies.

5   Exercises in PROJECT CREATION

5.1   Project FOUR-STROKE-ENGINE

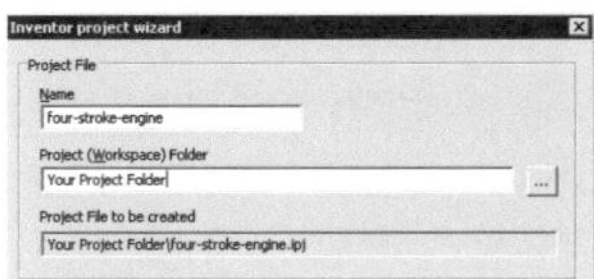

Now it is time to prepare a structured operation for our exercises. Now we will start a first individual project to make it easier to work properly later.

Tip: In the area of project folder you now choose a storage path for your project file. **Autodesk® Inventor® 2011** will connect all saving commands in context with this project to this storage path.

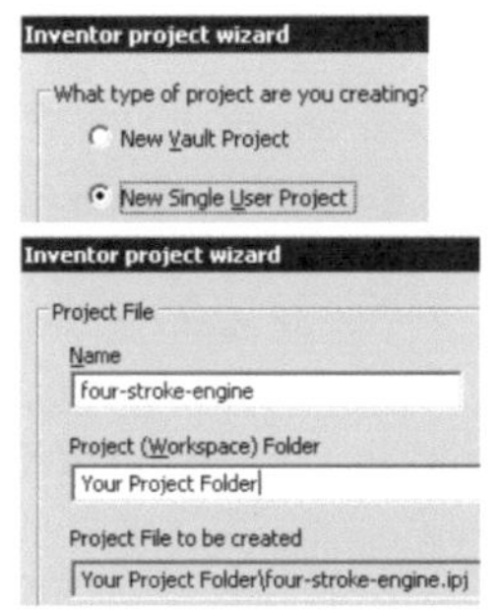

> ➤ Register: **Get Started**
> ➤ **Projects** > **New** > **Single User Project**
>
> ➤ Type in the name: **four-stroke-engine**
> ➤   Finish
>
> ➤ Double click with **LMC** on **four-stroke-engine**
> ➤   OK

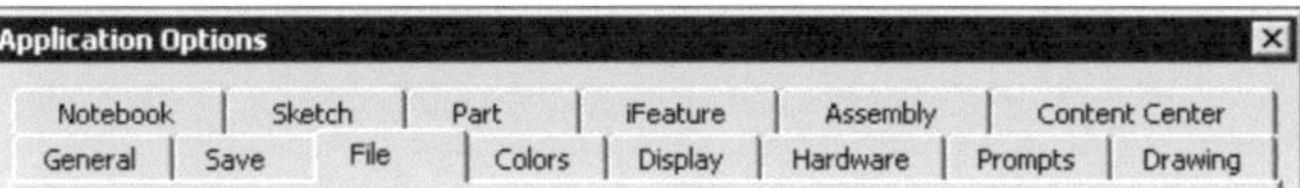

> ➤ Register: **Tools**
> ➤ **Application Options** > **File** > **Projects Folder**

Set the wanted future storage location for all files of the project **four-stroke-engine.** The project is now created and active.

Tip: Before starting work check for the correct settings in the areas **projects** and **project folder.** Create an individual folder for every new project. This regulatory measure will pay later.

---

6   SKETCHES & PARTS

6.1   Part VALVE
6.1.1   Basics of the part

The valves of our four-stroke-engine open and close the intake and exhaust manifolds of the cylinder head. They are controlled by the cams of the camshaft.

For our engine we need 8 valves as each combustion chamber has exactly one intake and one exhaust manifold.

6.1.2   Creating a sketch

For our first modeling exercise we chose a simple rotation part. The valve consists of a ☑ **2D Sketch** and a volume shape, which is created by 🛢 **Revolve.** Create a 🗇 **New Part** and 💾 **Save** it as **valve.ipt** in the project folder you created. The sketch area now automatically opens. In the area model (folder **Origin**) are the planes Y-Z, X-Y and X-Z as well as the axes X, Y and Z, which we also can make visible, however a use of these (for measurements and constraints) is not possible like this. First we have to use the command 🗐 **Project Geometry**.

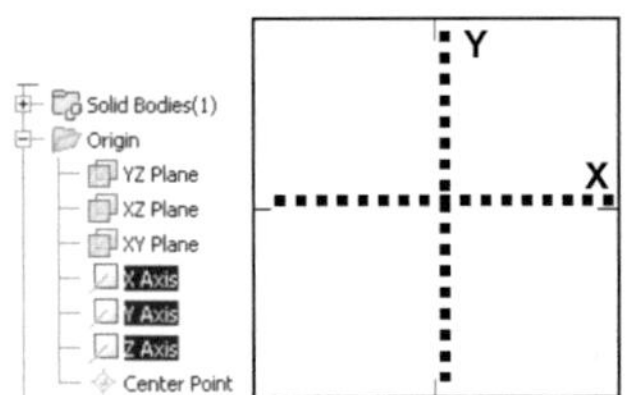

> 🗐 **Project Geometry** (X-Y-Z-axes)
> In the browser window (under **Origin**) you now choose the X-Y-Z-axes, then press the **ESC** key to finish the command.

Now the X-Y-Z-axes are active in the drawing area.

Tip: When you create a new sketch or edit a sketch you first use the command 🗐 **Project Geometry** to project the 3 main axes (X, Y, Z) into the coordinates origin of the sketch. Without this step you cannot use the 3 main axes.

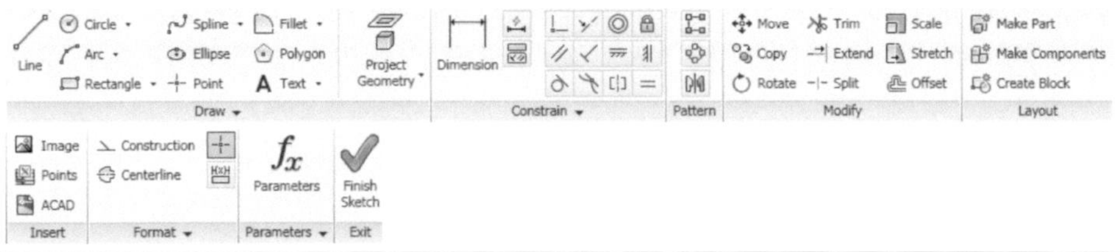

The command cluster **Draw** contains all basic drawing commands. Choose a command, set the first point with the **LMC** in the drawing window and set another the second point with renewed **LMC** at another spot of the chosen command. With the **ESC** key you end the respective command.

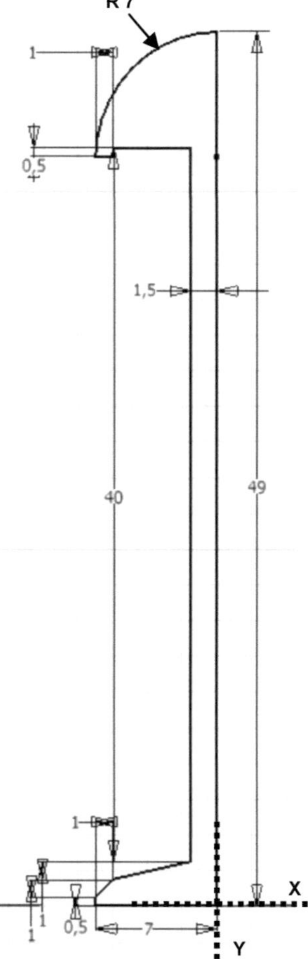

Draw the depicted outline. Use the various drawing commands of the command cluster **Draw**.

With the command ⊓ **Dimension** you can dimension the outline. Click **LMC** on the first dimension point, then **LMC** on the second dimension point and then with **LMC** on the spot at which the dimension should be placed.

Use the **Constrains** to set defined constraints among the lines.

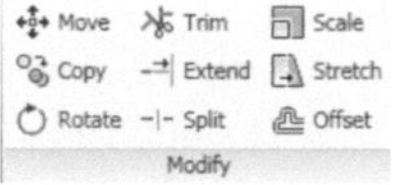

Use the commands in the area **Modify** to make changes on the lines.

When the sketches are complete finish them with following command:

✔ **Finish Sketch**.

Make sure that the lines with a length of 45 and 7 are placed on the shown axes (the Y-axis later will be our rotation axis for the volume shape). Use the command ⋎ **Collinear Constraint**, to place the lines onto the axis.

Tip: Quietly test the individual commands. In case of questions to the commands move the mouse pointer over the respective command and then press the **F1-key.** You will then be directed to a detailed explanation in the Help file.

6.1.3 Volume shapes by rotation

Our sketch is now complete. With the command ⊜ **Revolve** the existing sketch is transformed into a volume shape by rotation. As **Profile** you choose the just created area and as **axis** the Y-axis. Simple outline edges can also be used as rotation axis, it is just important that the area to be rotated is closed.

Should the outline not be accepted as area go back to the sketch (double click on the sketch in the model tree), then choose an outline, then **RMC** > **Close Loop**. Choose all adjoining lines one after the other until the outline is closed. Now, by using the command ⊜ **Revolve**, create following volume shape.

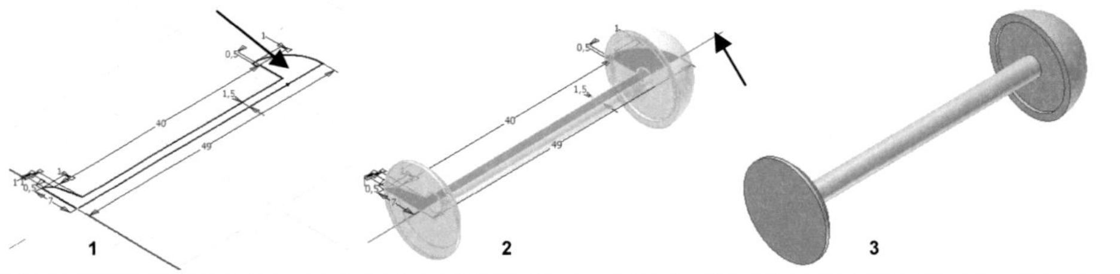

<table>
<tr><td align="center">1</td><td align="center">2</td><td align="center">3</td></tr>
</table>

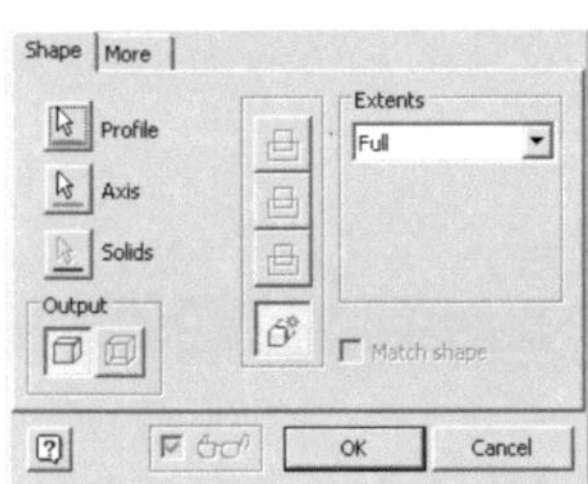

- ⊜ **Revolve**
- Profile: Area (picture 1)
- Axis: Y-axis (picture 2)
- Operation: Join
- Extents: Full (picture 3)
- OK

Tip: In the area **Output** you also can create a pure surface object. Should there be problems with the command ⊜ **Revolve** (it is possible that the sketch is not recognized as rotation area) re-open the sketch and close the outline. To do this click with **RMC** on one of the lines and choose the command **Close Loop.**

In the model tree window **Sketch1** was integrated in the command **Revolution1**.

With double click of **LMC** on **Sketch1** you get back to the respective sketch area and there you can make geometrical changes.

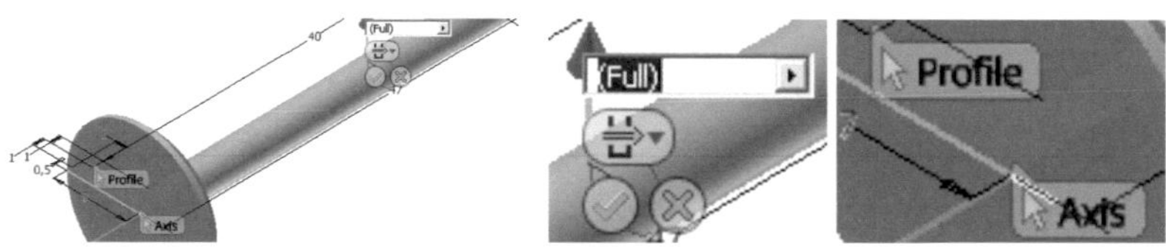

Also you get back to the command 🛞 **Revolve** by double clicking on **Revolution1** where you can change the created rotation if needed. Should it be necessary you also can correct your constructive steps here at any time. The values you also can edit directly in the sketch area. **valve.ipt** now is complete. 🖫 **Save** and ⊠ **Close** the file.

6.2   Part CRANKSHAFT PULLEY
6.2.1   Basics of the part

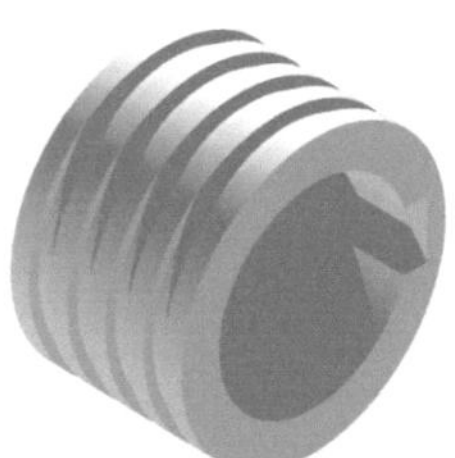

The pulley of the crankshaft transfers the rotary movement via a belt to the pulley of the camshaft.

As the rotation speed of the crankshaft has to be twice as fast as the rotation speed of the camshaft the pulley of the crankshaft only half the size in outer diameter as the one of the camshaft.

6.2.2   Create a basic sketch

The following part is also created via 🛞 **Revolve**. Create a ⊡ **New Part** and save it as **crankshaft-pulley.ipt** in the created project folder. 📽 **Project** the X-Y-Z-axes and draw the following outline.

Make sure it is located symmetrically to the Y-axis. The two distances 10 and 14 are in regards to the respective distance to the X-axis. Use the command ⌐ **Coincident**, to align the marked line symmetrically to the Y-axis.

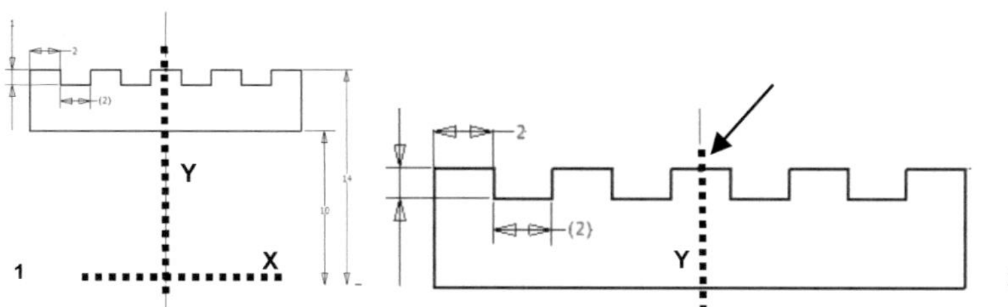

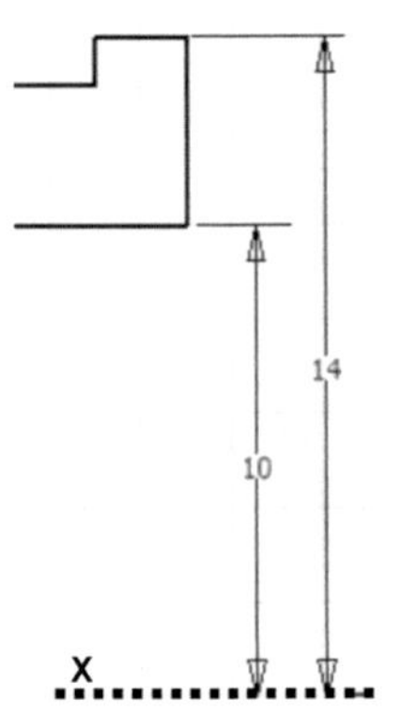

>   🖩 **Project Geometry** (X-Y-Z-axes)
> Draw the shown outline

Use the constraint └ **Coincident** to align the outline concentrically to the Y-axis.

> ✔ **Finish Sketch**

Tip: Choose the command └ **Coincident Constraint**, hover with the mouse pointer above the marked line in picture 2 to the center until a green dot shows up. Click this with **LMC** and then choose the Y-axis. The line automatically adjusts symmetrically to the Y-axis.

6.2.3   *Volume shape by rotation*

Now, by using the command 🔄 **Revolve**, create following volume shape.

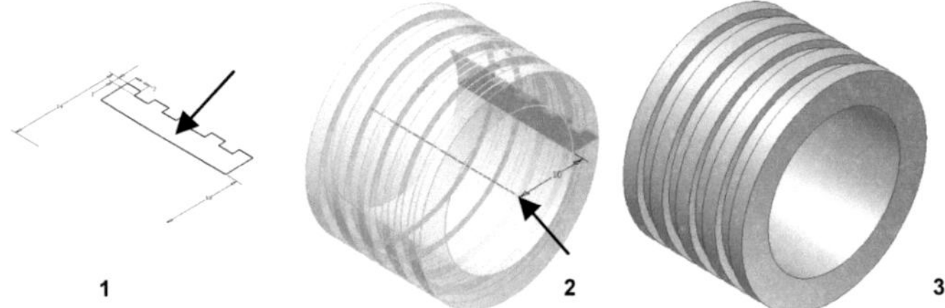

1   2   3

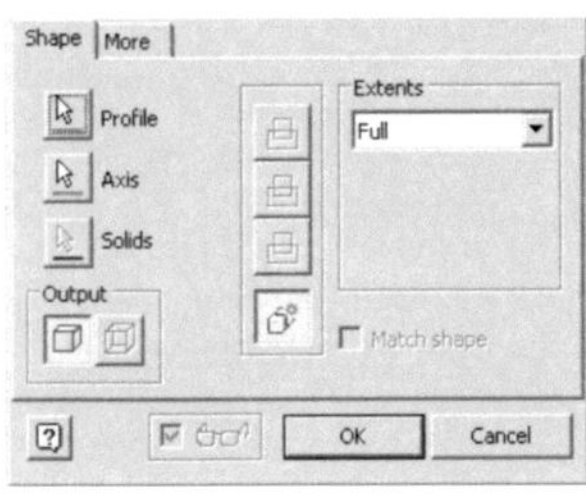

> 🔄 **Revolve**
> Profile: Area (picture 1)
> Axis: X-axis (picture 2)
> Operation: Join
> Extents: Full (picture 3)
>   OK

6.2.4 Creating a fitting key groove

To create the notch for the fitting key groove we need a new sketch on one of the two faces of the rotation shape. Choose one of the two faces and then use the command ✐ **2D Sketch**. Use the command ▣ **View Face** to align the sketch. Then you draw the shown rectangle. Use the command ∟ **Coincident Constraint**, to symmetrically align the rectangle to the Y-axis and then use the command ▤ **Extrude**, to create a volume shape. Make sure to use the procedure **Cut** to remove the material.

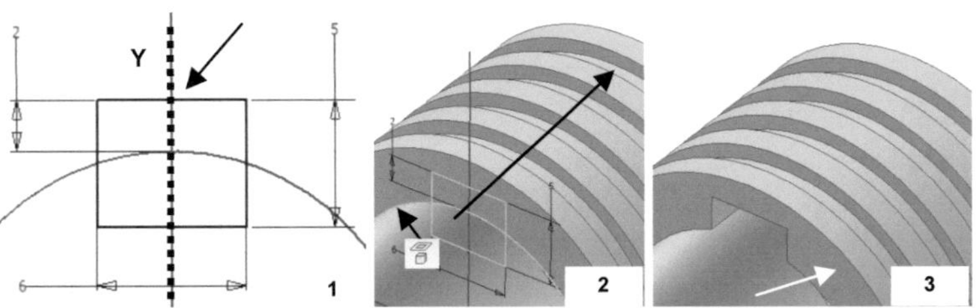

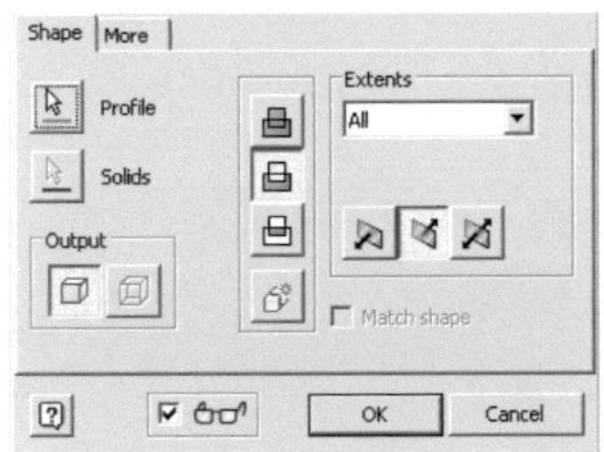

> ✐ **2D Sketch** (on marked area picture 3)
> ▤ **Project Geometry** (X-Y-Z-axes and the marked edge picture 2)
> ▣ **View Face** (on the marked face picture 3)
> Draw shown rectangle
> ✔ **Finish Sketch**

> ▤ **Extrude**
> Profile: Rectangle
> Operation: Cut
> Extents: All
> Direction: Showed picture 2
> OK

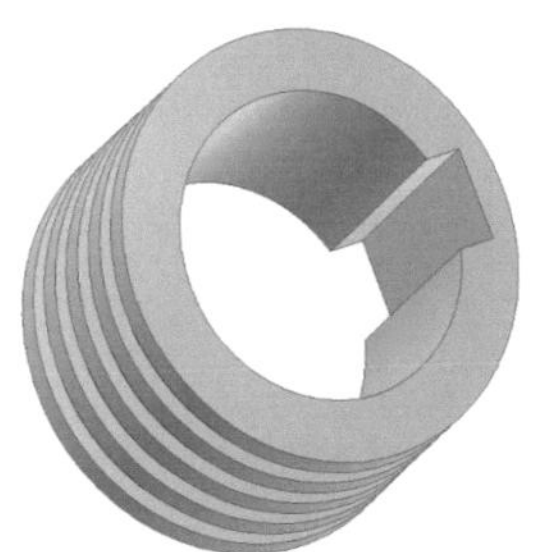

Tip: The rectangle (5 x 5) has to get aligned symmetrically to the Y-axis (picture 1) with the command ∟ **Coincident Constraint**.

---

crankshaft-pulley.ipt is now complete. ▣ **Save** and ✖ **Close** the file.

6.3 Part CAMSHAFT-PULLEY
6.3.1 Basics of the part

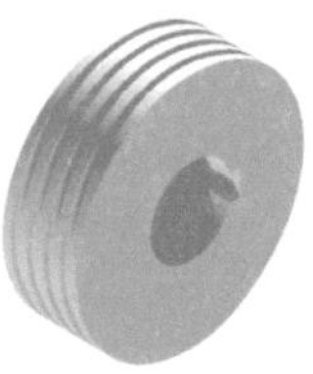

The pulley of the camshaft transfers the rotation movement of the crankshaft onto the camshaft.

It is driven by the timing belt. To create the pulley of the camshaft we use the geometrical analogy of the pulley to the crankshaft.

In a four-stroke-engine the camshaft has to turn half as fast as the crankshaft. This means we need a gear ratio of the rotation speed from camshaft to crankshaft of 1:2. We solve this by an outside diameter ratio camshaft to crankshaft of 1:2.

Our camshaft outer diameter therefore has to be twice as high as the one of the crankshaft. For this we simply change the height measurement of the outer diameter of the existing pulley of the crankshaft and save the new construction as pulley of the camshaft.

6.3.2 Deriving from existing constructions

📂 **Open** the file **crankshaft-pulley.ipt** and via command 💾 **Save as** create the file **camshaft-pulley.ipt**.

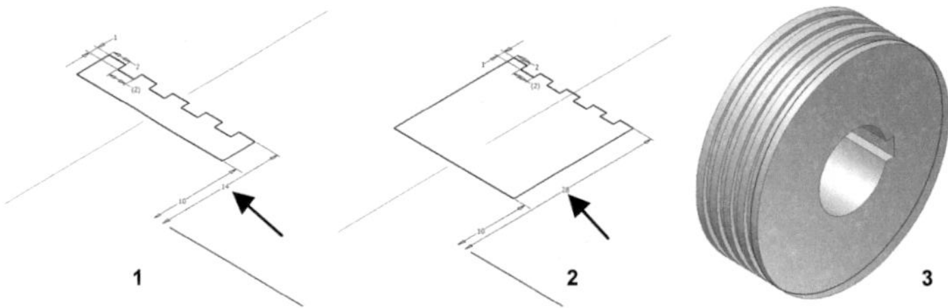

1   2   3

> Open **Sketch1** under 3D-command **Revolution1**
> Change the distance 14 by double clicking to 28 
> ✔️
>
> ✔️ **Finish Sketch**

So in the existing **Sketch1** we simple change the height measurement 14 to 28.

The rotary element **Revolution1** automatically changes (if not use the command 🔄 **Update**). **camshaft-pulley.ipt** now is complete. 💾 **Save** and ✖ **Close** the file.

6.4   Part CYLINDER SLEEVE
6.4.1   Basics of the part

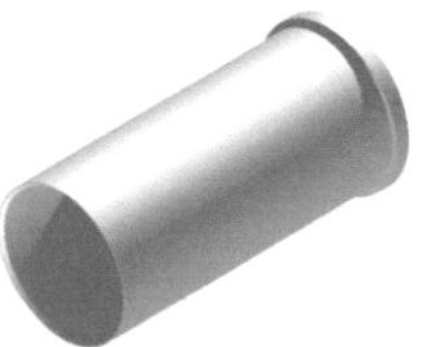

Cylinder blocks (engine blocks) are made of aluminum or iron casting. To enable a better thermal capacity of the cylinder walls cylinder sleeves are inserted. The sleeve (cylinder sleeve) emulates the surface for the piston and is inserted into the cylinder block.

6.4.2   Base body by rotation

Create a 🗀 **new part** and 💾 **save** it as **sleeve.ipt.** Then you draw a new outline and afterwards create a volume shape with the command 🌐 **Revolve**.

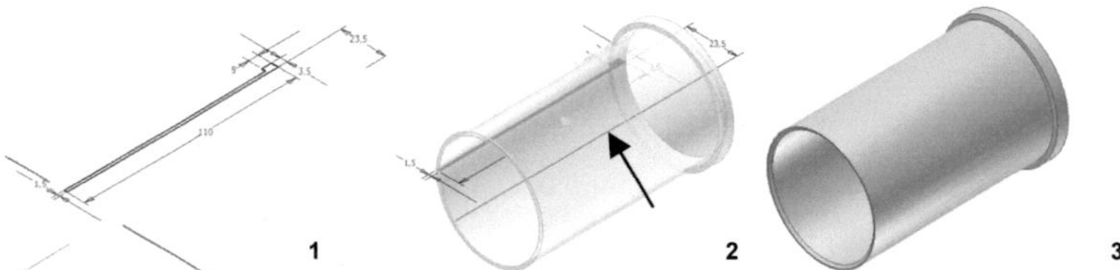

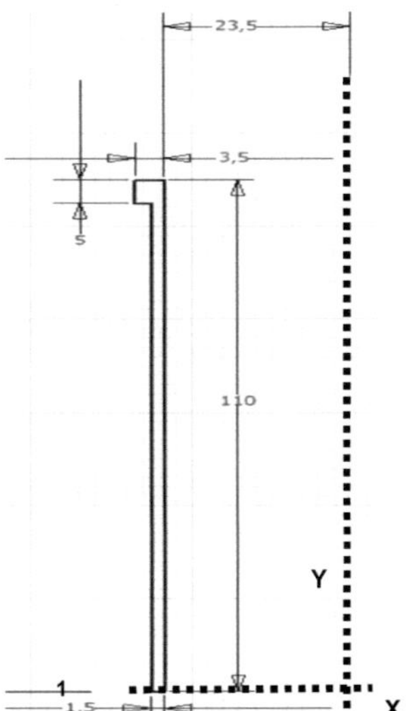

> ⬚ **Project Geometry** (X-Y-Z-axes)
> Draw shown outline
> ✔ **Finish Sketch**

> 🌐 **Revolve**
> Profile: Area (picture 1)
> Axis: Y-axis (picture 2)
> Operation: Join
> Extents: Full (picture 3)
> [OK]

6.4.3   Chamfer of the inside edge

After the volume shape is created we now provide it with a ◪ **Chamfer**. Use the procedure **Distance and Angle** and the references inside edge of the liner and area of the lower face (picture 2).

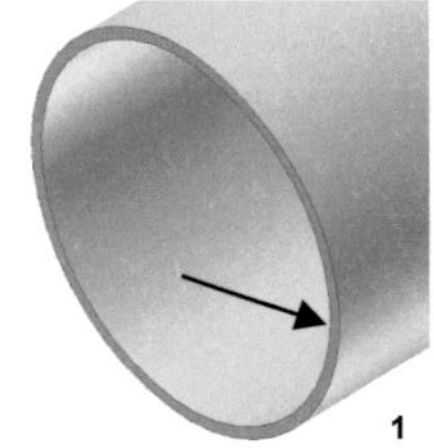 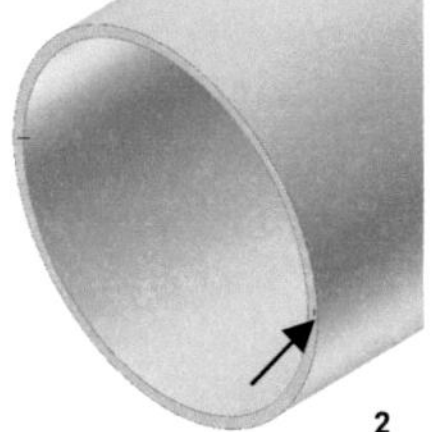 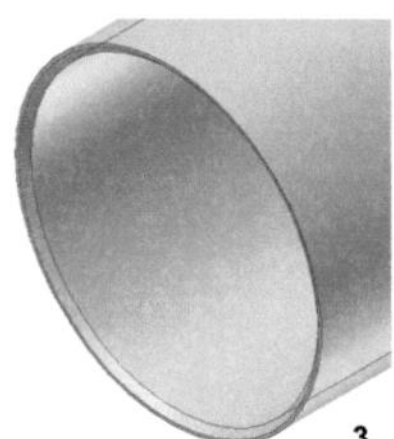

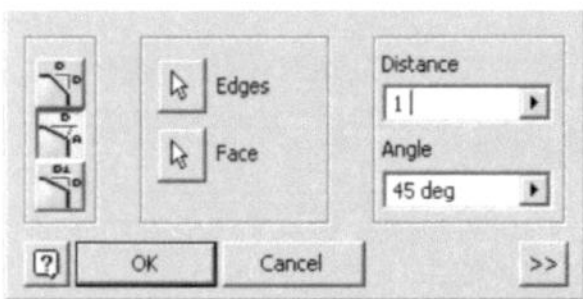

> ◪ **Chamfer**
> Type: Distance and Angle
> Edges: Lower inside edge of the cylinder (picture 1)
> Face: Lower face of the cylinder (picture 2)
> Distance: 1
> Angle: 45°
> OK

Tip: In the command ◪ **Chamfer** are next to the option **Distance** two more options for bevel adjustment: **Distance + Angle** and **Two Distances**. As needed you can work here with angles or clearances.

sleeve.ipt now is complete. 🖬 **Save** and ☒ **Close** the file.

6.5   Part SPARK PLUG
6.5.1   Basics of the part

The ignition voltage generated by the ignition system is transferred to the spark plug via the ignition lead.

The voltage jumps from the rotor electrode over to the distributor cap electrode and thereby creates the ignition spark which then ignites the fuel/air mixture.

Our spark plug consists of two threads (located on the electrode), one insulator and a hexagon for bolting.

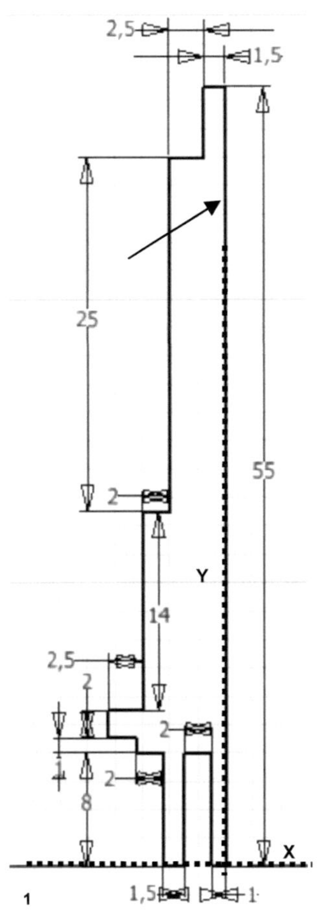

6.5.2   Basic body by rotation

Create a 🗁 **New Part** and 💾 **Save** it as **sparkplug.ipt** 🗗 **Project** the X-Y-Z-axes and draw following outline.

Make sure that the marked line of the outline is placed on the Y-axis as it will later be our rotary axis (use the command: 🗡 **Collinear Constraint**). Afterwards use 🖰 **Revolve** to turn a volume shape.

➢ 🗗 **Project Geometry** (X-Y-Z- axes)
➢ Draw shown outline
➢ ✔ **Finish Sketch**

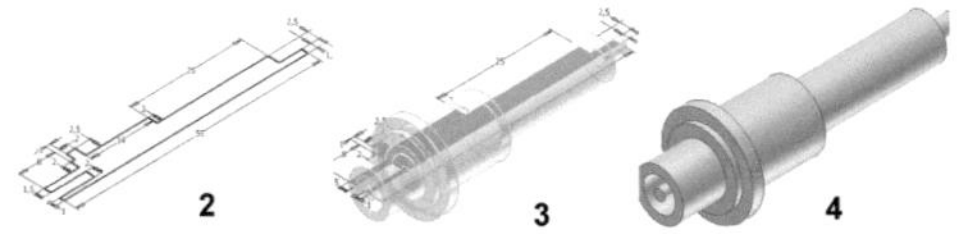

➢ 🖰 **Revolve**
➢ Profile: Area
➢ Axis: Y-axis
➢ Operation: Join
➢ Extents: Full
➢   OK

6.5.3   Rounding the insulator

With the command 🖰 **Fillet** we place the three marked roundings.

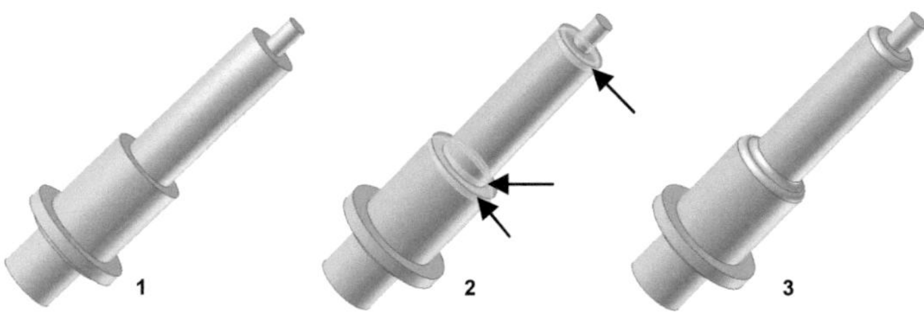

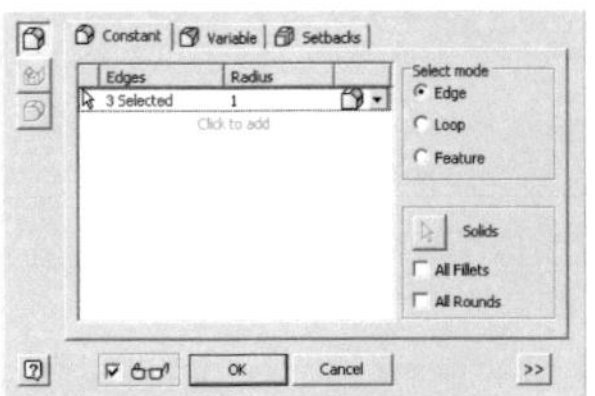

> 🖉 *Fillet*
> Edges: 3 marked edges
> Radius: 1
> ☐ OK

Tip: The command 🖉 *Fillet* contains several options. You can choose between *Constant*, *Variable* or *Setbacks* with various sub-items.

6.5.4   Thread to screw in the sparkplugs

With the command 🗐 *Thread* we now will create two different threads on the existing cylindrical areas.

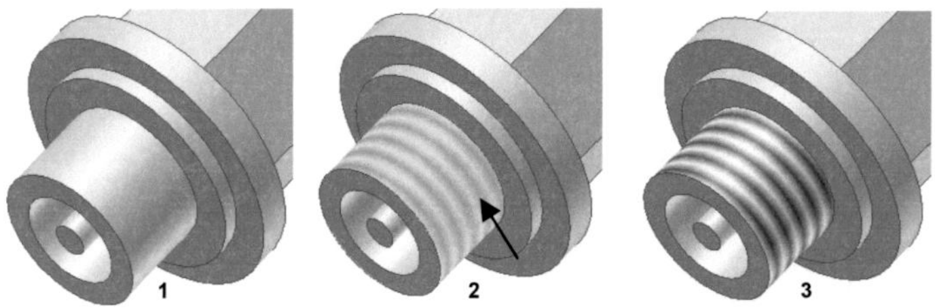

> 🗐 *Thread*
> Area: Marked cylinder face (picture 2)
> Thread Type: ISO Metric profile
> Size: 9 - Right hand
> Designation: M9 x 1.25
> Class: 6g
> ☐ OK

Note: If you are using *inch* units, please transform the metric ones to inch system.

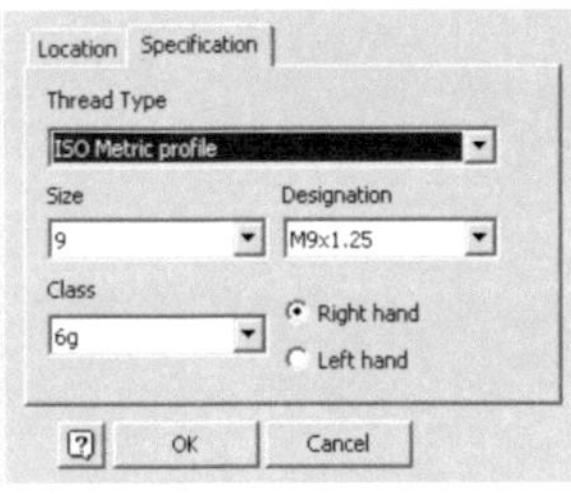

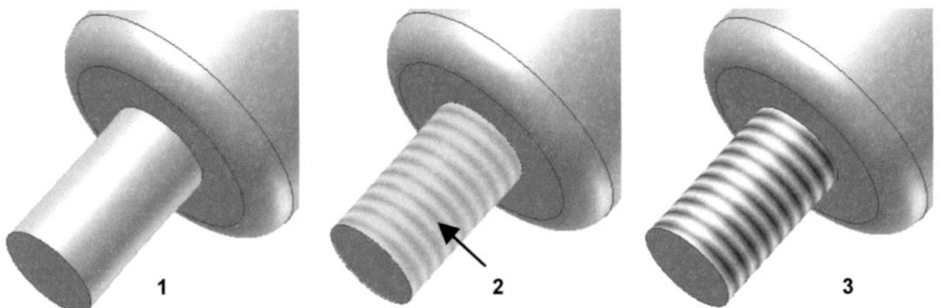

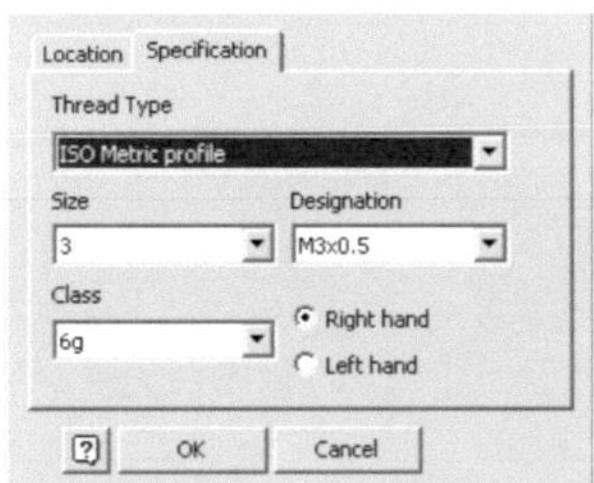

> ⬛ ***Thread***
> ➤ Area: Marked cylinder face (picture 2)
> ➤ Thread Type: ISO Metric profile
> ➤ Size: 3 - Right hand
> ➤ Designation: M3 x 0.5
> ➤ Class: 6g
> ➤   OK

Note: If you are using ***inch*** units, please transform the metric ones to inch system.

6.5.5   *Chamfer of the thread*

The side of the thread which later is screwed into the cylinder head shall get a small ◁ ***Chamfer*** for easier mounting.

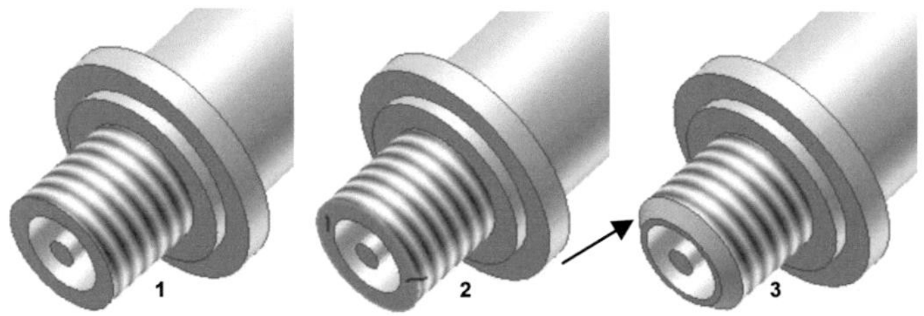

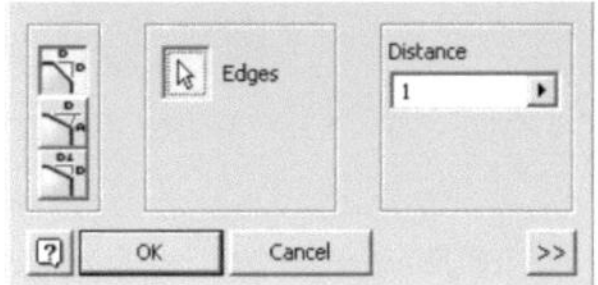

> ➤ ◁ ***Chamfer***
> ➤ Edges: Marked edge (picture 3)
> ➤ Distance: 1
> ➤   OK

6.5.6   *Creating a polygon*

Afterwards we will create the still missing polygon for screwing in the sparkplugs with the command ▯ ***Extrude***.

To do this place a new ▱ ***2D Sketch*** on the marked area in picture 1 (use the ***F7*** key to slice a volume shape to the active sketch), ⬛ ***Project*** the X-Y-Z-axes, draw the shown ⬠ ***Polygon*** and create an ▯ ***Extrude***.

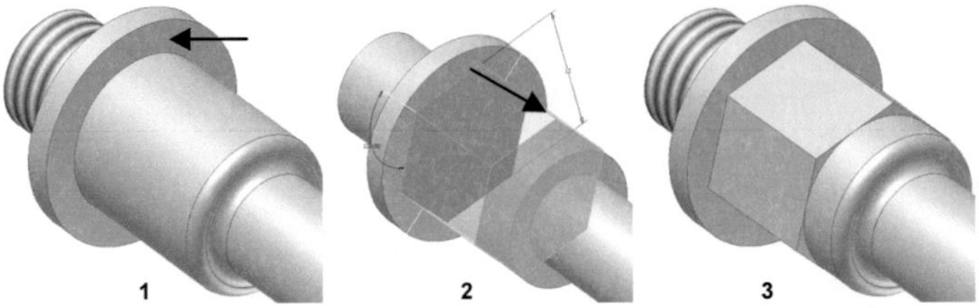

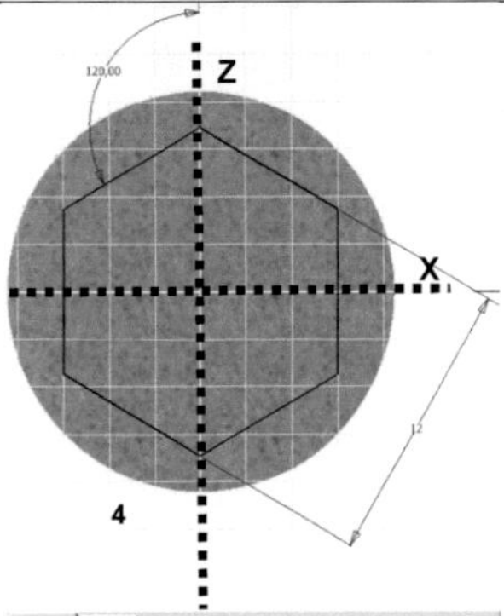

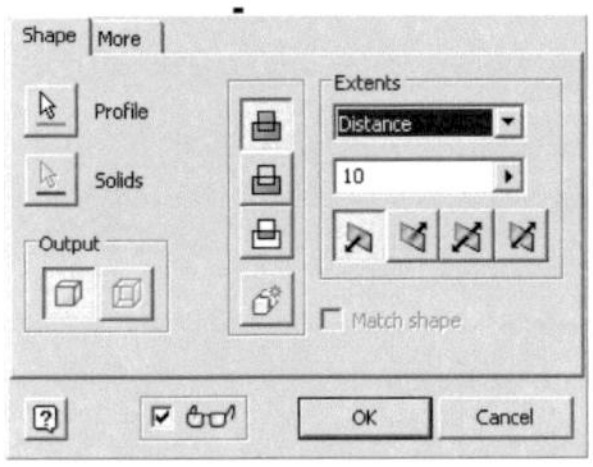

> 🖋 **2D Sketch** (on marked area picture 1)
> Slice the sketch with **F7**
> 🖨 **Project Geometry** (X-Y-Z-axes)
> Draw the shown 6-point-polygon (line distance 12)

> ✔ **Finish Sketch**

> 🔲 **Extrude**
> Profile: Polygon
> Operation: Join
> Extents: Distance 10
> Direction: shown (picture 2)
> OK

sparkplug.ipt now is complete. 💾 **Save** and ✖ **Close** the file.

6.6   Part PISTON
6.6.1   Basics of the part

A piston is used for linear movable sealing of the combustion chamber and for power transfer of the burning expansion onto the piston rod.

The piston consists of a hollow body, the grooves for the piston seal rings, a hole for the wrist pin and two notches.

The pistons for a piston engine are made of aluminum casting alloys or forged of cast iron.

6.6.2   Basic body by rotation

Create a ⬚ **New Part** and 🖫 **Save** it as **piston.ipt.** ⬚ **Project** the X-Y-Z-axes and then draw the shown outline. Make sure that the marked line of the outline is located on the Y-axis as this will be our rotation axis. Then create the volume shape with the command ⬚ **Revolve**.

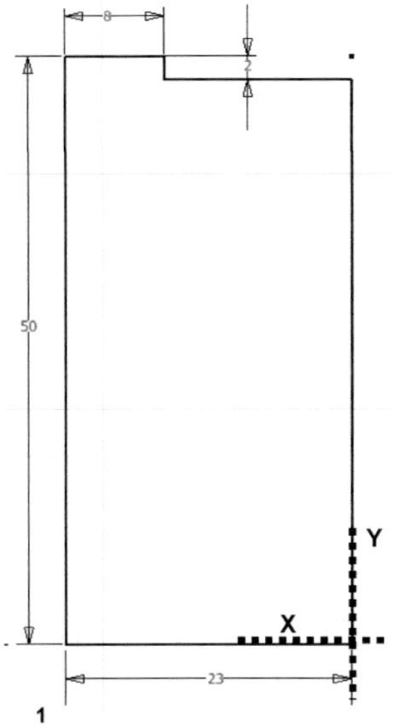

> ➤ ⬚ **Project Geometry** (X-Y-Z-axes)
> ➤ Draw shown outline

> ➤ ✔ **Finish Sketch**

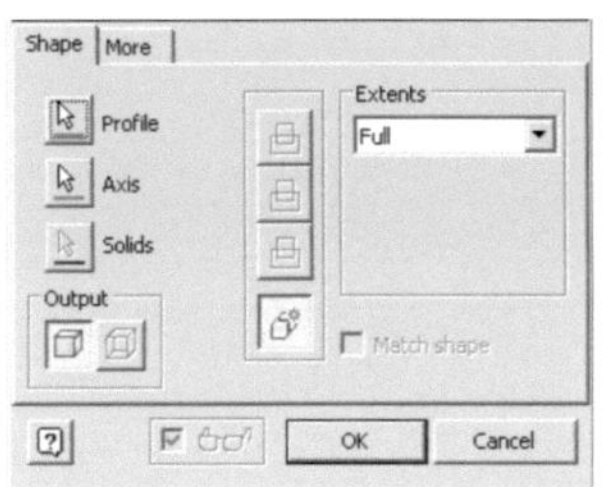

> ➤ ⬚ **Revolve**
> ➤ Profile: Area (picture 1)
> ➤ Axis: Y-axis
> ➤ Operation: Join
> ➤ Extents: Full
> ➤ [OK]

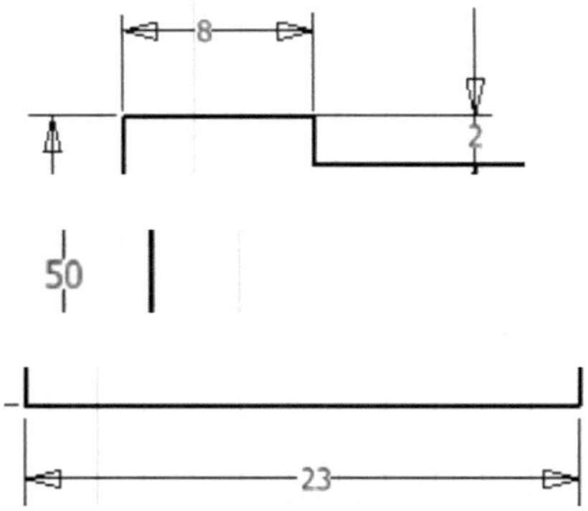

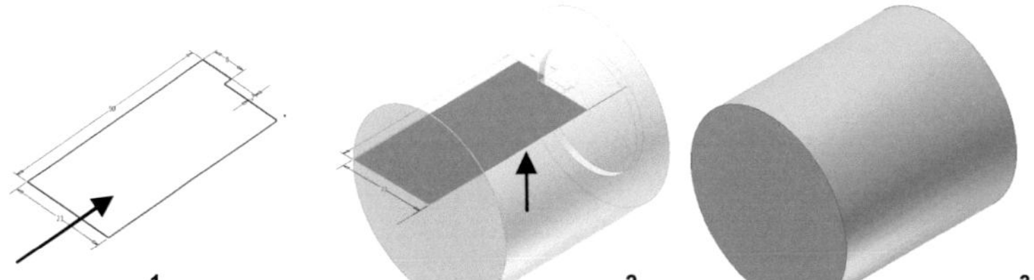

6.6.3   Rounding the upper piston area

With the command ◌ *Fillet* both edges are rounded.

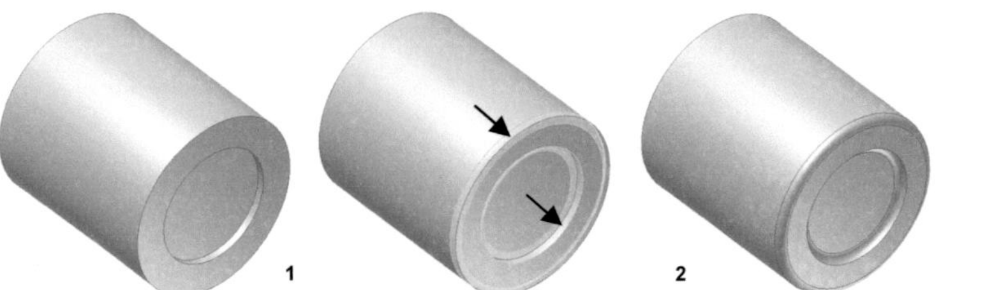

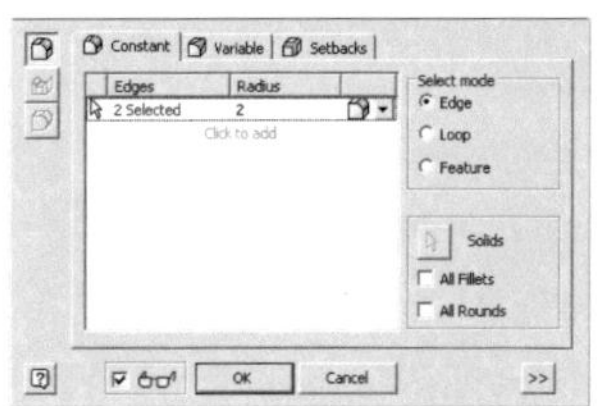

> ◌ *Fillet*
> Edges: 2 marked edges (pictrue 2)
> Radius: 2
> ☐ OK

6.6.4   Notches for the piston rings

In our exercise book no piston rings are used, however, we still will plan the notches for them.

Create a new ✎ *2D Sketch* on the X-Y-plane, slice it with *F7* and draw the shown outline. Use the command ⠿ *Rectangular Pattern*.

Then create with the command ⊟ *Revolve* the material cut of the 3 notches.

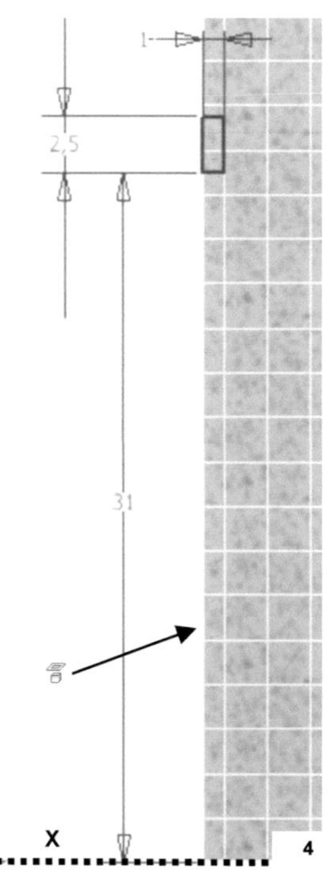

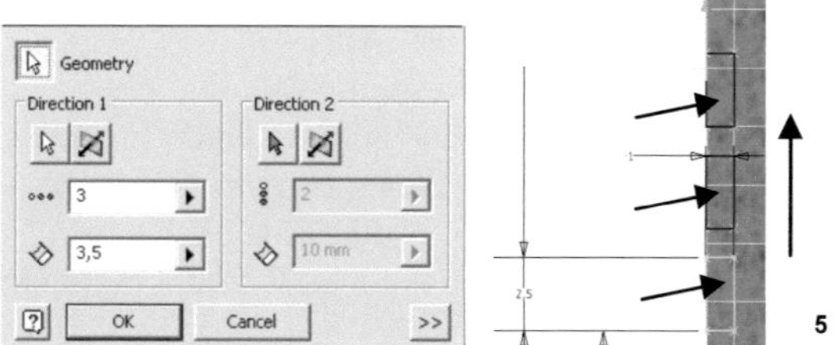

- ➢ ✍ **2D Sketch** (on X-Y- plane)
- ➢ Slice the sketch with **F7**
- ➢ ⛭ **Project Geometry** (X-Y-Z-axes and the marked edge of the piston in picture 4)
- ➢ Draw shown rectangle (picture 4)

- ➢ ⊞ **Rectangular Pattern**
- ➢ Elements: Rectangle
- ➢ Direction: Y-axis (showed in picture 5)
- ➢ Quantity: 3
- ➢ Distance: 3.5
- ➢ [OK]

- ➢ ✔ **Finish Sketch**

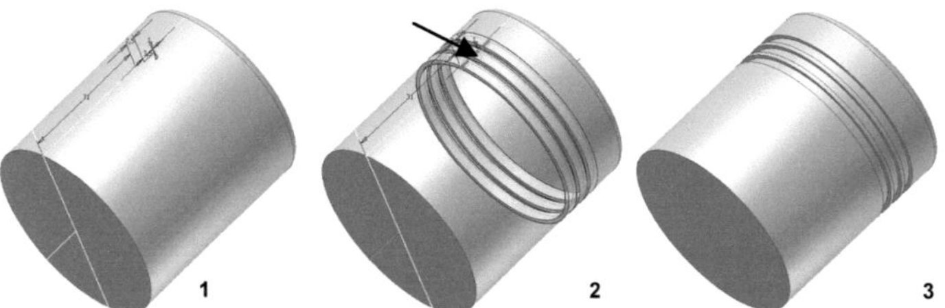

- ➢ ⬒ **Revolve**
- ➢ Profile: 3 rectangles
- ➢ Axis: Y-axis
- ➢ Operation: Cut
- ➢ Extents: Full
- ➢ [OK]

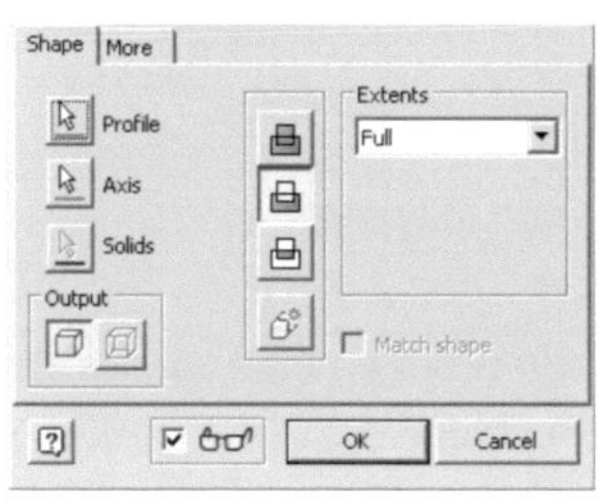

6.6.5 Outline for the wrist pin

Now we need 2 notches for the piston pin. Create a new ✏ **2D Sketch** on the X-Y-plane, slice it with **F7** and draw the shown outline. Use the command ▯◁ **Mirror**, to create the second rectangle, then remove the two areas with the command ▯ **Extrude** from the existing material.

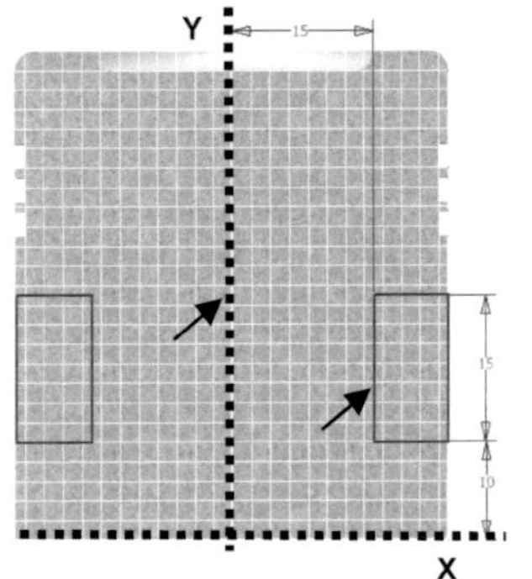

> ➢ ✏ **2D Sketch** (on X-Y-plane)
> ➢ 🖉 **Project Geometry** (X-Y-Z-axes)
> ➢ Slice the sketch with **F7**
> ➢ Draw the right rectangle according to sketch
>
> ➢ ▯◁ **Mirror**
> ➢ Elements: Right rectangle
> ➢ Axis: Y-axis
> ➢ [Apply]
> ➢ [OK]
>
> ➢ ✔ **Finish Sketch**

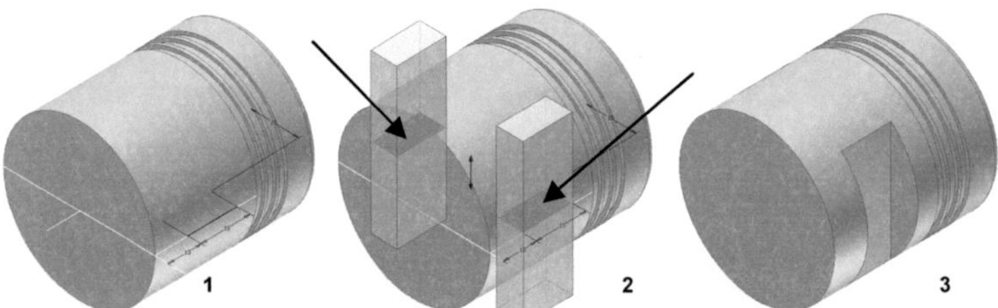

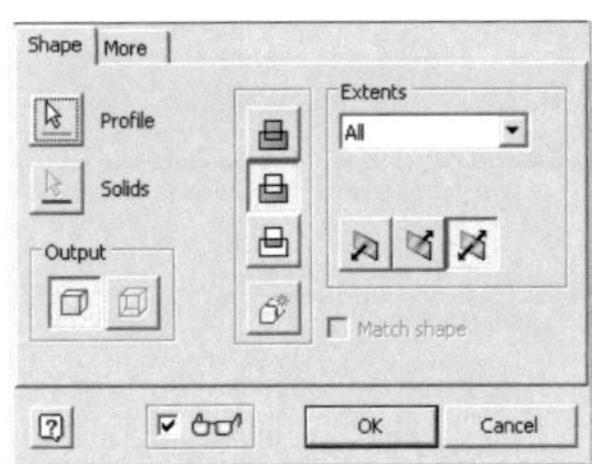

> ➢ ▯ **Extrude**
> ➢ Profile: Both rectangles
> ➢ Operation: Cut
> ➢ Extents: All
> ➢ Direction: Symmetrical
> ➢ [OK]

Then we set the hole for the piston pin. Create a new ✏ **2D Sketch** on one of the two newly created faces of the piston. 🖉 **Project** all needed lines and draw a circle.

The circle center should be located on the projected Y- axis. Then subtract the resulting volume shape via the command ▯ **Extrude**.

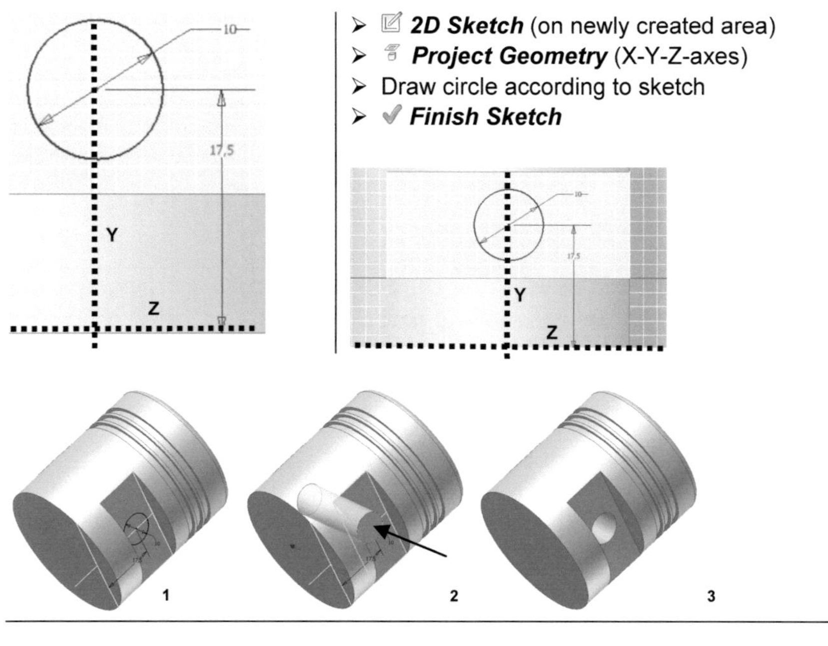

- ➤ ☑ *2D Sketch* (on newly created area)
- ➤ ☞ *Project Geometry* (X-Y-Z-axes)
- ➤ Draw circle according to sketch
- ➤ ✔ *Finish Sketch*

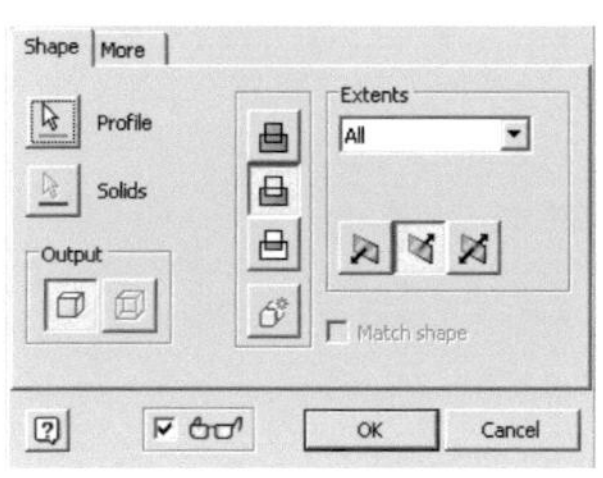

- ➤ ⬚ *Extrude*
- ➤ Profile: Circle
- ➤ Operation: Cut
- ➤ Extents: All
- ➤ Direction: Shown (picture 2)
- ➤ OK

6.6.6 Hollow body by wall thickness

The outer outline of the piston now is complete. We now use the command ⬚ *Shell*, to adjust the piston from the inside to the outer outline with a specified wall thickness.

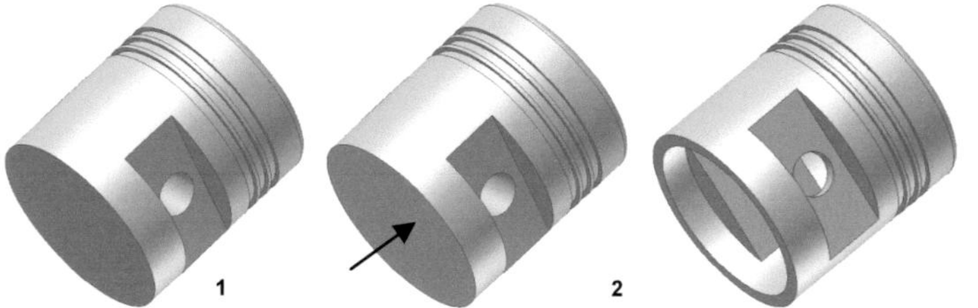

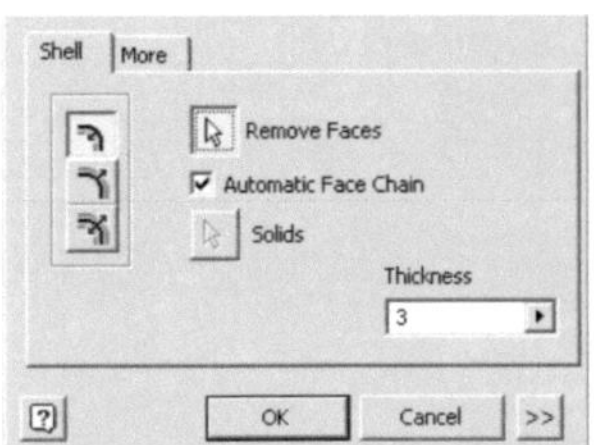

> ☐ **Shell**
> Remove Faces: Marked face (picture 2)
> Thickness: 3
> Automatic Face Chain: Yes
> Direction: Within
> ☐ OK

Tip: The appointed direction determines whether material is added or removed. The program automatically searches for all adjoining areas and this way forms an outline adjusted to the object. Under **Further Options** you have the possibility to choose the marginal conditions for this operation.

6.6.7 Shifting of work steps within the model tree

It is time to correct a small blemish. The program also recognizes the previously executed material slice for the pin hole as volume shape, which also receives a wall thickness.

Therefore we now have material to remove at the position where later our wrist pin will be located. This is done via a simple trick. In the model tree we move the command **Shell1** above the command **Extrusion2** (Drag and Drop).

The program now recalculates the newly created outline of the piston.

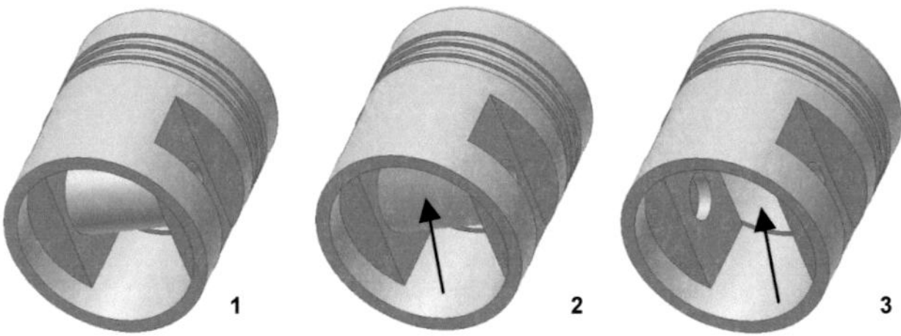

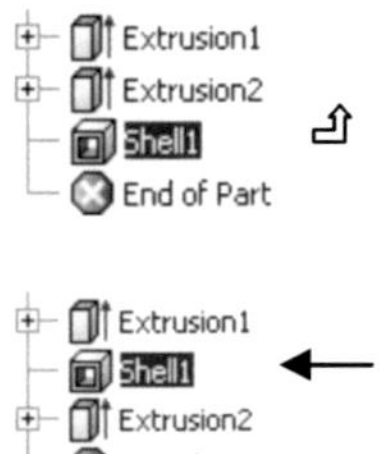

> Open the model tree
> Press the element **Shell1** with **LMC** and keep it pressed, then drag above **Extrusion2**

piston.ipt now is complete. ☐ **Save** and ✕ **Close** the file.

6.7   Part bottom of the piston rod
6.7.1   Basics of the part

The piston rod changes the linear movement of the piston into a linear oscillating axis movement. Often the piston rod is first made of a single part and then specifically split. Both parts then fit exactly and the breakage point no longer is visible after mounting.

Our piston rod consists of topside and a bottom side. First we make the bottom side to then construct the topside from it.

Create a ▣ *New Part* and ▣ *Save* it as ***bottom-side-piston-rod.ipt*** in the project folder. Then ▯ *Project* the X-Y-Z-axes and draw the outline. Afterwards ▯ *Extrude* symmetrically.

6.7.2   Volume shape by extrusion

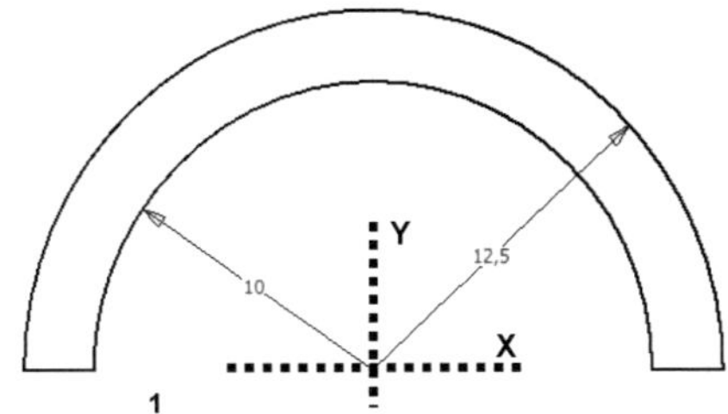

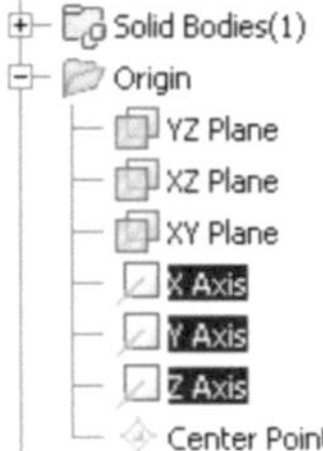

➢ ▯ *Project Geometry* (X-Y-Z-axes)
➢ ⌒ *Arc* (Center Point)

Draw two 180° circular arcs with the command ⌒ *Arc* (Center Point) and dimension them (R1= 10, R2 = 12.5). Then close the two open ends with two lines.

➢ ✔ *Finish Sketch*

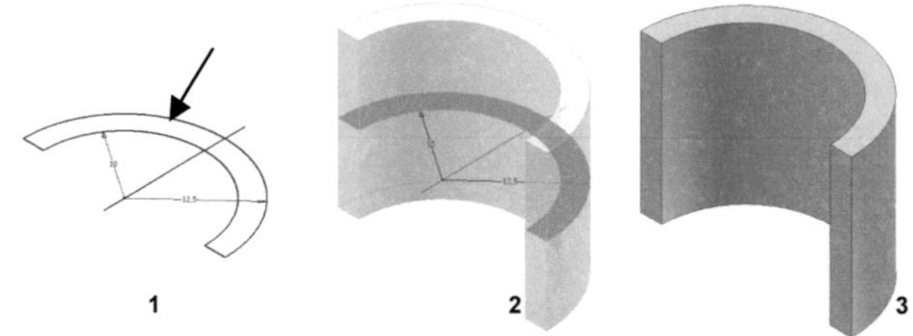

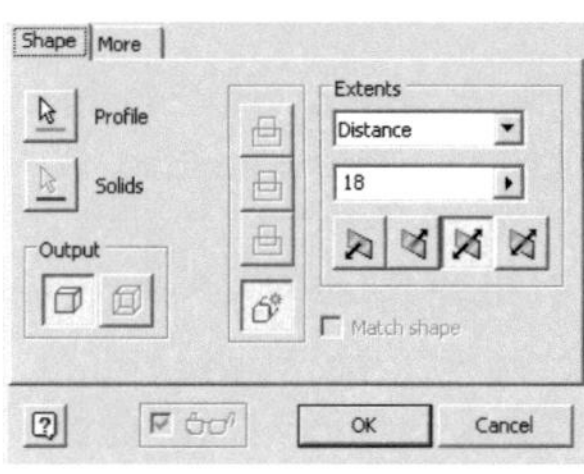

> 🗍 **Extrude**
> Profile: Marked area (picture 1)
> Operation: Join
> Extents: Distance 18
> Direction: Symmetrical
> [OK]

6.7.3 Grounding clips for screw connection

The two parts of the piston rod later are to be screwed together. For this we need two clamps with one hole each. On one of the two joint faces we will create a new 🗍 **2D Sketch,** then 🗍 **Project** the X-Y-Z-axes and draw all necessary edges and the two circles. Use the Z- axis to 🗍 **Mirror**. Afterwards we create a cylinder by using 🗍 **Extrude**.

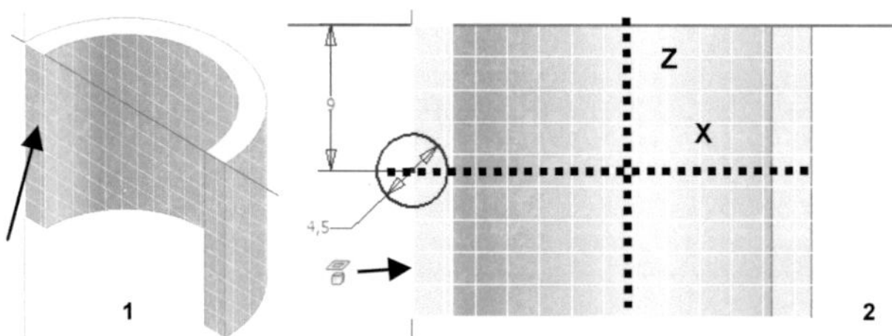

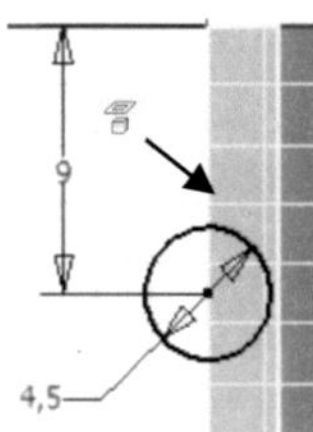

> 🗍 **2D Sketch** (on marked area picture 1)
> 🗍 **Project Geometry** (X-Y-Z-axes and the marked outer edge on the left showed in picture 2)
> Draw shown circle (D= 4.5)
> ✔ **Finish Sketch**

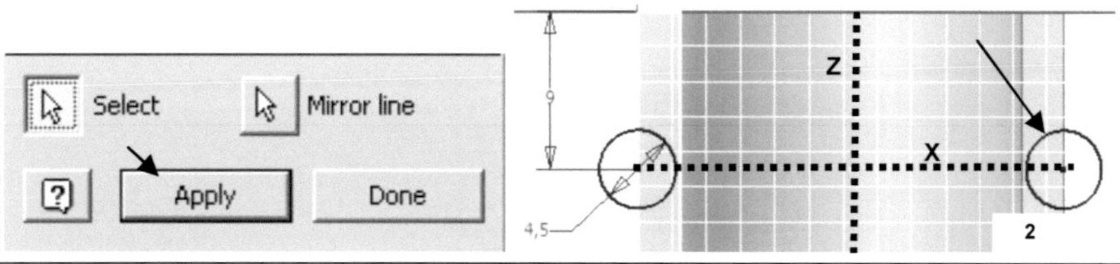

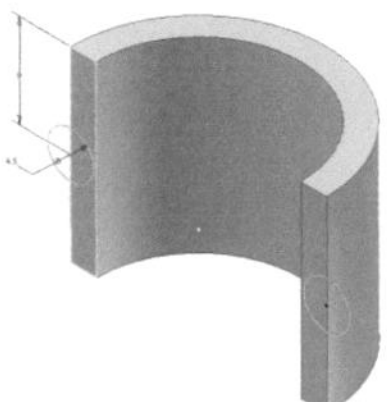

> ⫴ **Mirror**
> Select: Left circle (picture 2)
> Mirror line: Z-axis
> Apply
> Done
>
> ✓ **Finish Sketch**

Tip: Without the interim certificate with the command [Apply] **Apply** the mirroring possibly might not be executed properly.

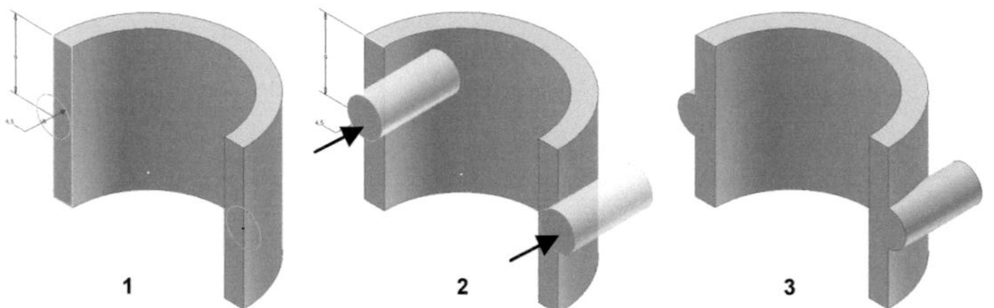

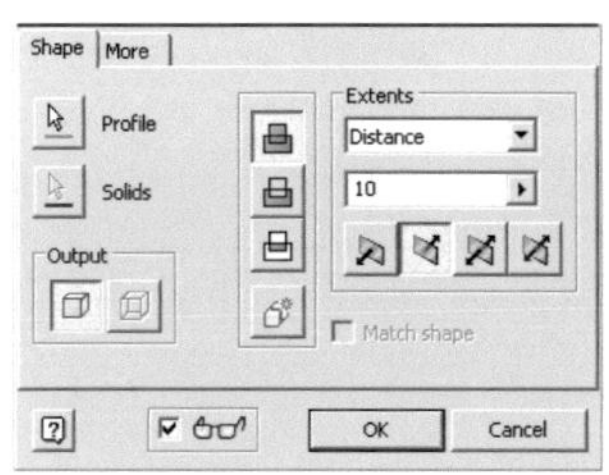

> ⬒ **Extrude**
> Profile: Both circle areas
> Operation: Join
> Extents: Distance 10
> Direction: Shown (picture 2)
> OK

Afterwards the two clips have to get a hole through which later the screws are placed.

Create a new ✎ **2D Sketch** on one of the two end faces, ⬚ **Project** the two existing cylinder edges, draw the two circles and then make a material slice with the command ⬚ **Extrusion**.

Make sure that the two new circles are ◎ **Concentic** to the two projected cylinder edges.

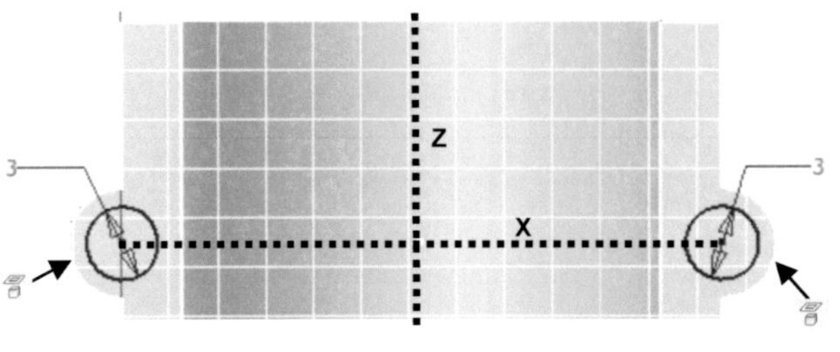

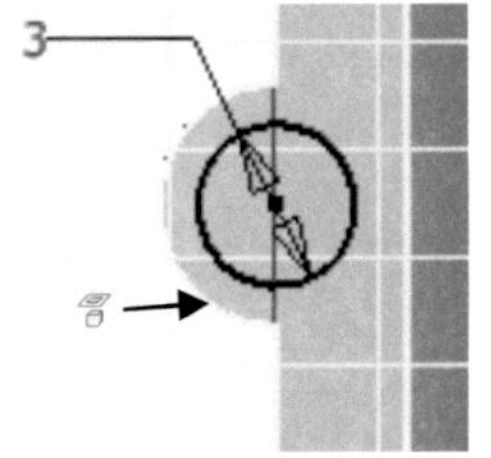

> ☑ **2D Sketch** (on one of the joint faces)
> ☞ **Project Geometry** (marked edges)
> Draw 2 circles (D= 3) concentic to the projected cylinder edges
> ✔ **Finish Sketch**

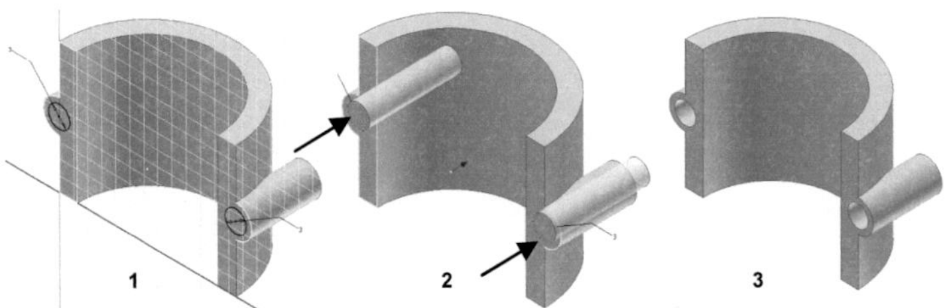

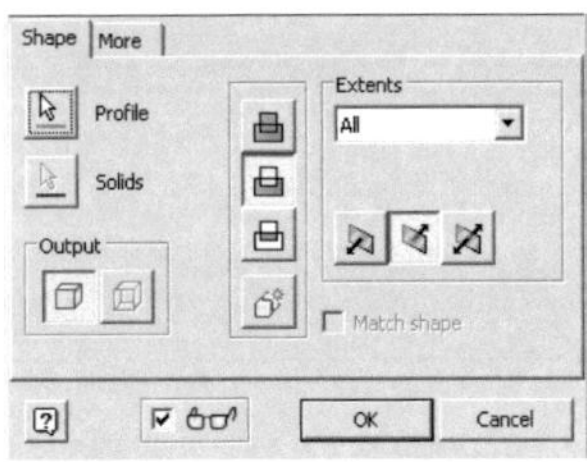

> ⬛ **Extrude**
> Profile: Both circle areas
> Operation: Cut
> Extents: All
> Direction: Shown (picture 2)
> [OK]

6.7.4   Chamfer & Fillet

For the finishing touch we will edit the part with the commands ⬤ **Chamfer** and ⬤ **Fillet**.

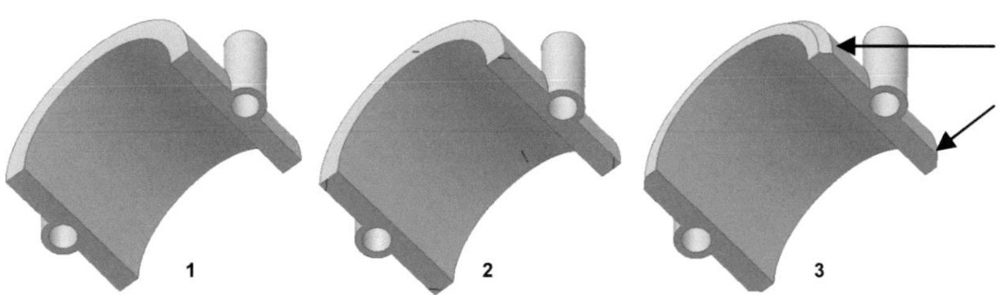

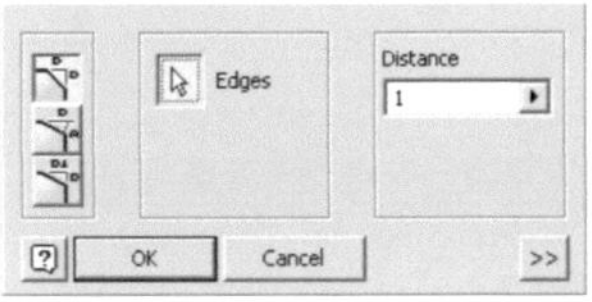

- ➤ **Chamfer**
- ➤ Edges: 2 marked edges
- ➤ Distance: 1
- ➤ OK

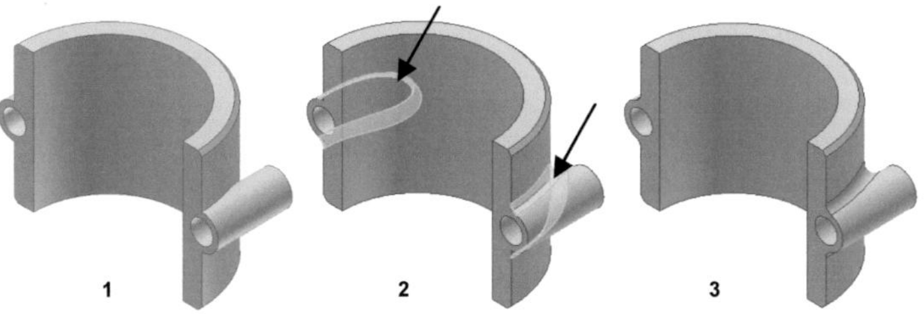

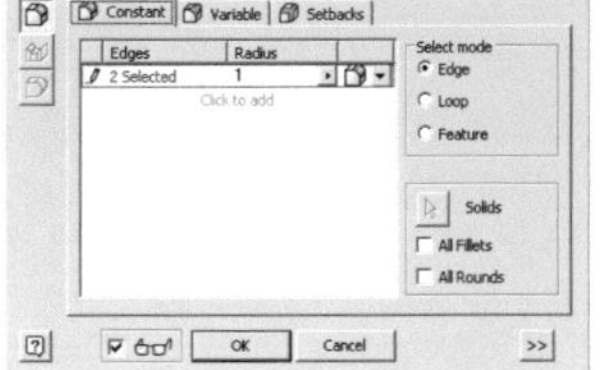

- ➤ **Fillet**
- ➤ Edges: 2 marked edges (pictrue 2)
- ➤ Radius: 1
- ➤ OK

bottom-side-piston-rod.ipt now is complete. **Save** and **Close** the file.

6.8   Part TOPSIDE PISTON ROD
6.8.1   Basics of the part

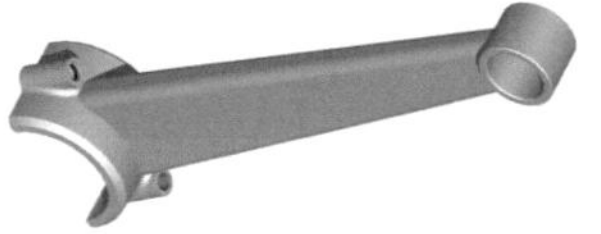

Our piston rod topside consists of a connecting- rod shank, a connecting- rod end and a bearing cup.

As the **bottom side piston rod** is geometrically identical to the **topside piston rod** we use the existing construction to make the task easier.

Open the file *bottom-side-piston-rod.ipt* and *Save* it as *topside-piston-rod.ipt* in your project folder. Create a new *Plane* parallel to the marked area (picture 1). To do this choose the command, click on the area and assign 12.5 in the field *Dimension* (picture 2). Should the plane be drifted into the wrong direction simply change the algebraic sign in the (input) screen.

6.8.2   Bottom connecting rod shaft

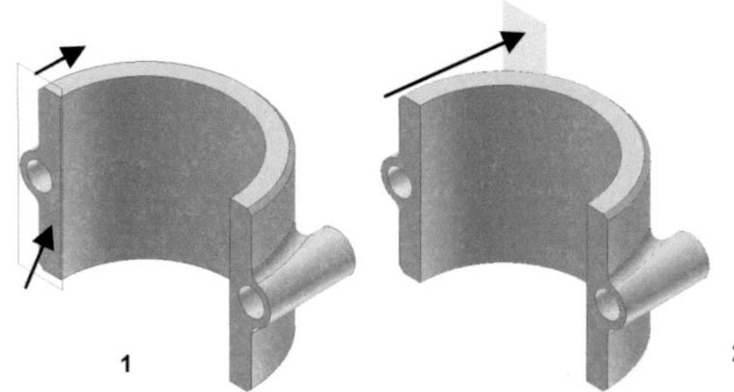

> *Plane*
> Click the marked face (picture 1)
> Dimension: 12.5
> ✓

Tip: By double click with *LMC* in the model tree on a created plane you can change the specified distance at any time.

The distance value of the plane can be positive or negative. Simply test in what direction it goes and correct it with a negative algebraic sign if needed. Create a new *2D Sketch* on the created plane and draw the shown rectangle.

Adjust the sides of the rectangle centered on the prior projected axes. Then create the volume shape in the direction of the existing body by using the command *Extrude*. Choose the procedure *To Next* for this.

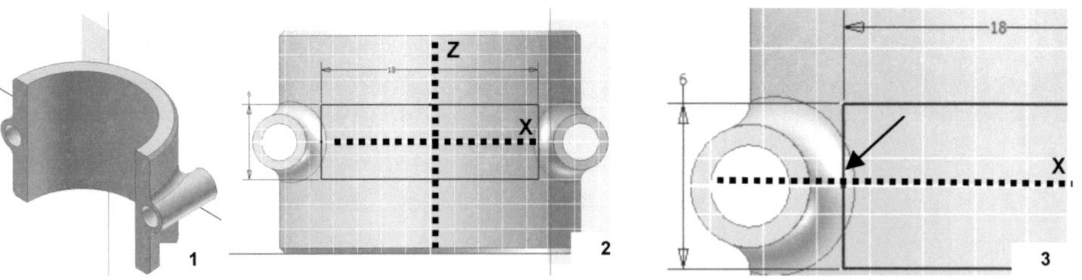

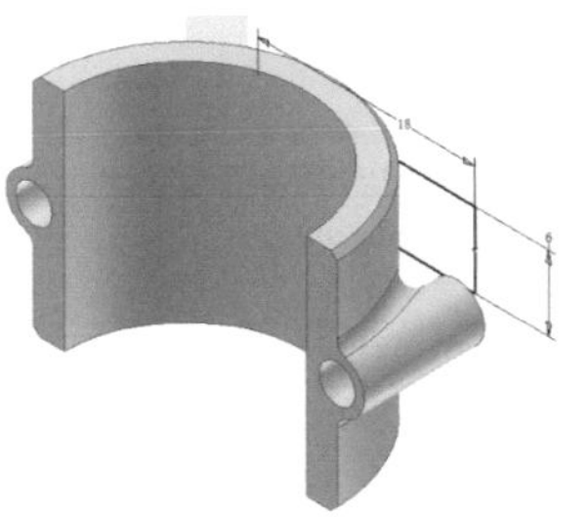

> ✐ **2D Sketch** (on new plane)
> ☞ **Project Geometry** (X-Y-Z-axes)

Draw a rectangle (18 x 6) and adjust the line meridians of the rectangle to the projected axes with the command ∟ **Coincident Constraint**.

> ✔ **Finish Sketch**

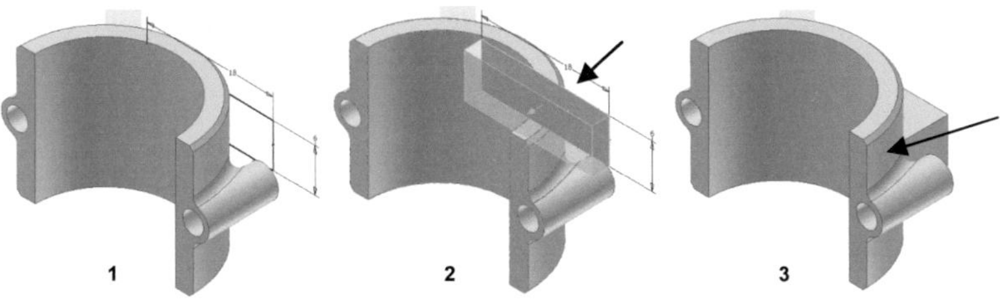

1   2   3

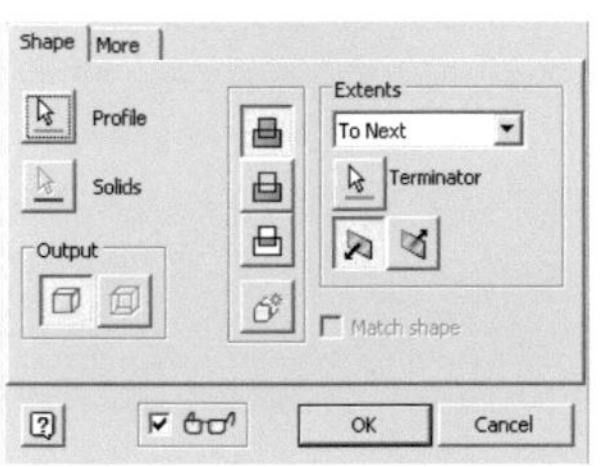

> ▯ **Extrude**
> Profile: Rectangle
> Operation: Join
> Extents: To Next
> Terminator: Marked face in picture 3
> Direction: Shown (picture 2)
>   OK

6.8.3   Top connecting rod shaft

The bottom connecting rod shaft now cleanly mates with the bottom cup. Now we can construct the remaining connection rod bridge with the command ▽ **Loft**. For this we need a new plane and two new sketches.

Create a ✐ **2D Sketch** on the marked area shown in picture 1, project all edges of the marked face and then finish the sketch again. After that you create a new ▱ **Plane**, parallel to the area shown in picture 1, create another ✐ **2D Sketch** on it and draw the rectangle (10 x 5) shown in picture 3.

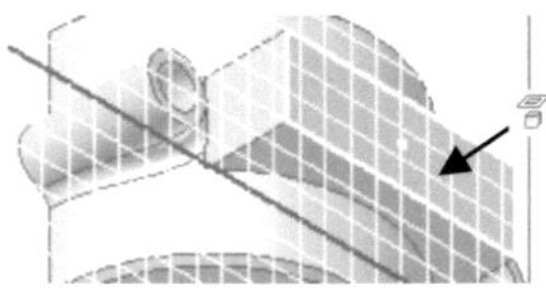

> ✐ **2D Sketch** (on marked area)
> ☞ **Project Geometry** (4 edges of the marked area)
> ✔ **Finish Sketch**

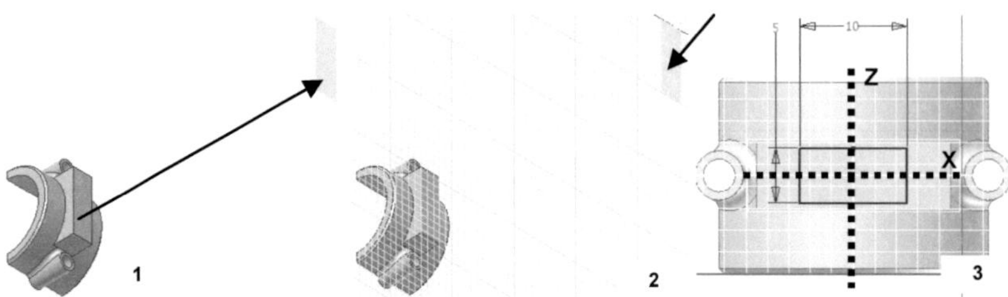

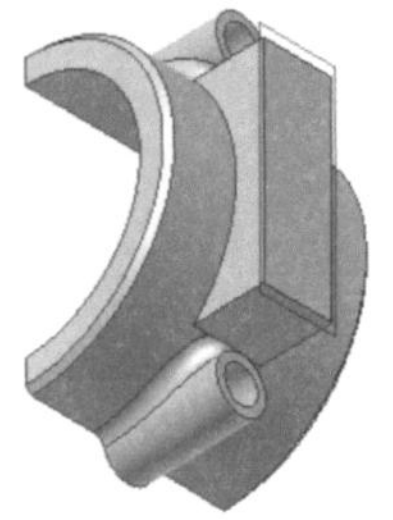

> ⬚ *Plane*
> Click on the marked area (picture 1), keep **LMC** pressed and drag the mouse in the shown direction (picture 1)
> Distance: 87
> ✔

> ✎ *2D Sketch* (on new plane showed in picture 2)
> ☷ *Project Geometry* (X-Y-Z-axes)

Draw a rectangle (5 x 10) and align to the projected axes by └ *Coincident Constraint.*

> ✔ *Finish Sketch*

Use the command ⬡ *Loft*, to create a topside connection rod shaft between the two rectangles.

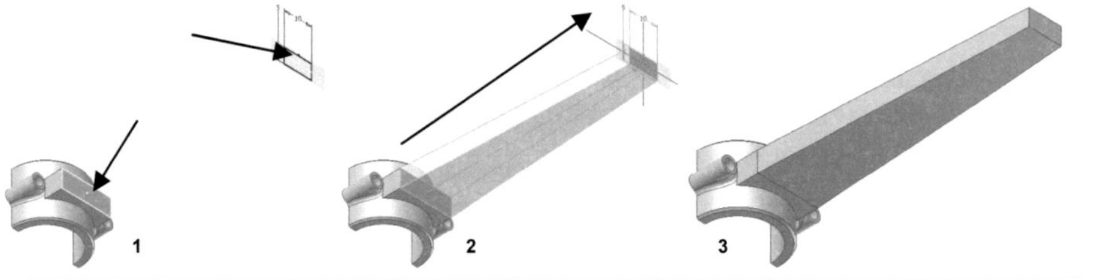

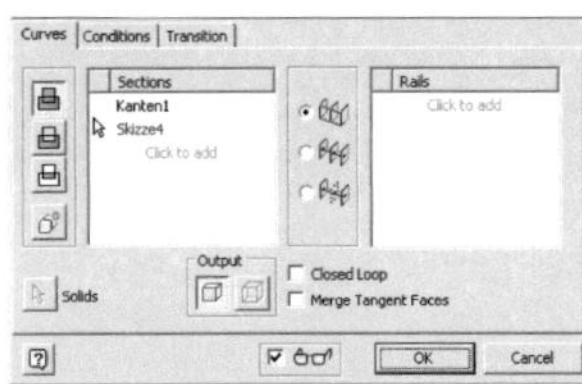

> ⬡ *Loft*
> Operation: Join
> Sections: Both rectangles (picture 1)
> Rails: None
> Output: Volume shape
> ⬚ OK

Tip: The command ⬡ **Loft** offers other options. You can direct the elements along a **path** (progress guide), create **area elements** (Output), or edit the connection resp. starting points in the areas **Conditions** and **Transition**.

6.8.4 Connecting rod end for the bolt guide

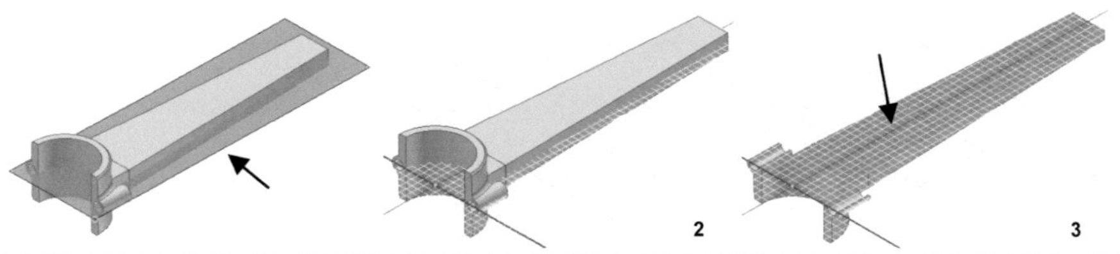

For the connecting rod end we create a new 🖉 **2D Sketch** on the X-Y-plane, draw the shown circle and then create a cylinder with the command 🗋 **Extrude**.

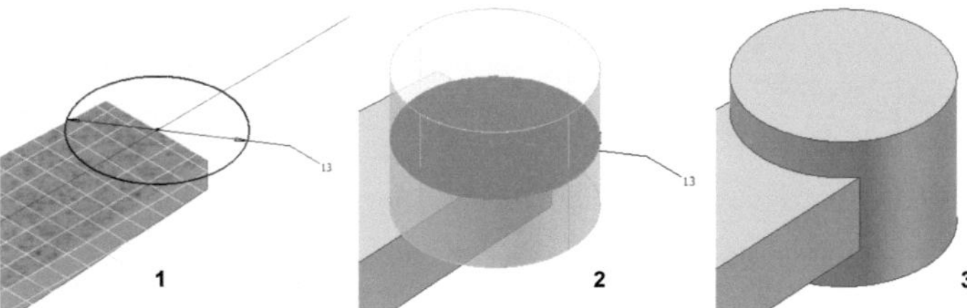

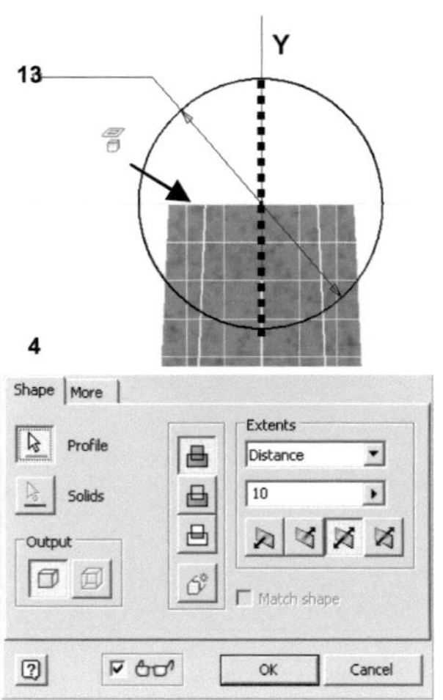

> ➤ **Project Geometry** (X-Y-Z-axes and marked edge picture 4)
> ➤ Slice the sketch with **F7**

Draw a circle with D= 13 centrically to the Y-axis and the projected line.

> ➤ ✔ **Finish Sketch**

> ➤ 🗋 **Extrude**
> ➤ Profile: Circle area
> ➤ Operation: Join
> ➤ Extents: Distance (10)
> ➤ Direction: Symmetrical
> ➤ ☐ OK

With the command ⬡ **Hole** we create the guide for the pin. Create a new ✎ **2D Sketch** on the marked face of the created cylinder, roject the marked geometry of the cylinder and place a ✛ **Point**. Afterwards create a ⬡ **Through-hole**.

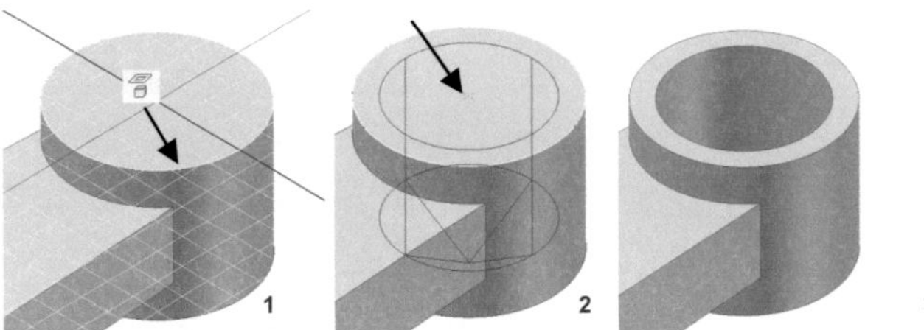

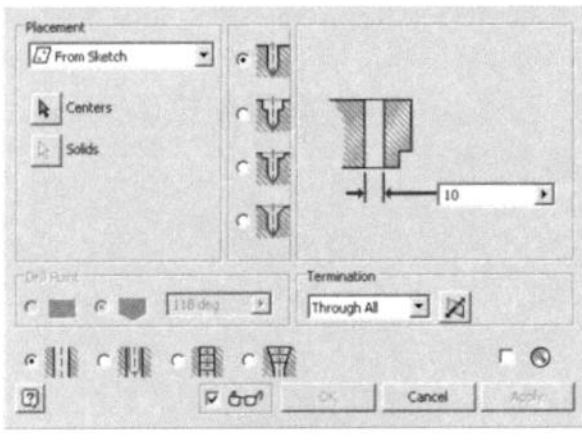

> ✎ **2D Sketch** (on one of the faces of the created cylinder)
> ✐ **Project Geometry** (marked cylinder edge picture 1)
> ✛ **Point** draw in the center of the created circle
> ✔ **Finish Sketch**

> ⬡ **Hole**
> Placement: From Sketch
> Type: Through All
> Diameter: 10
> ⬦ OK

6.8.5　Fillet & Chamfer

Afterwards we round the side edges of the connecting rod shaft and provide the connecting rod end with 4 ◈ **Chamfers**.

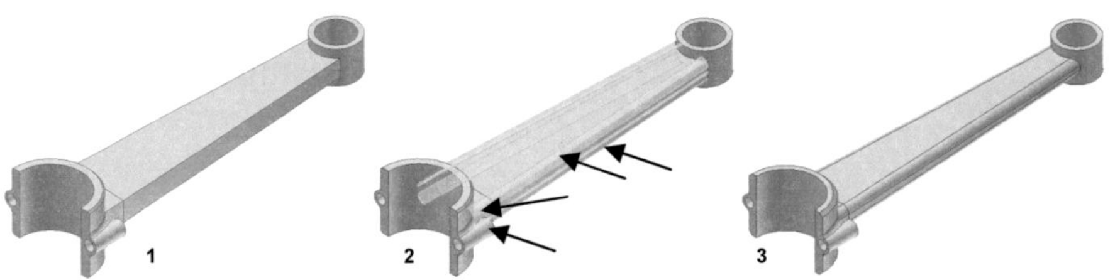

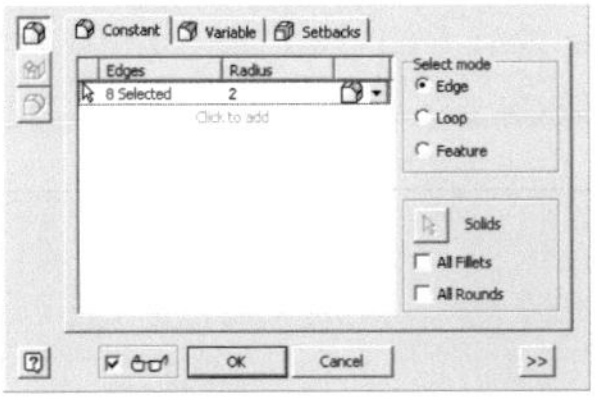

- ***Fillet***
- Type: Constant
- Edges: 8 edges of the connecting rod shaft (both sides)
- Radius: 2
- OK

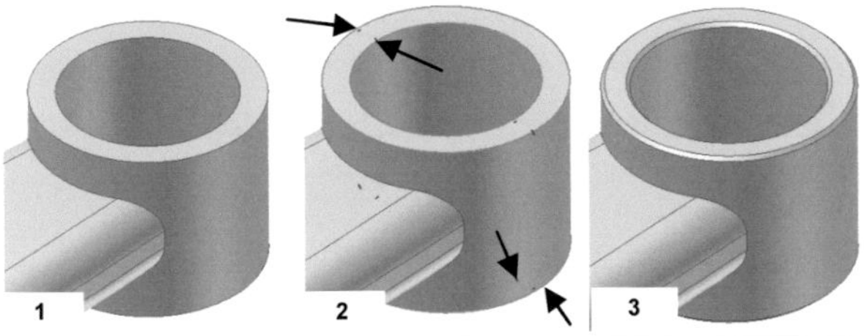

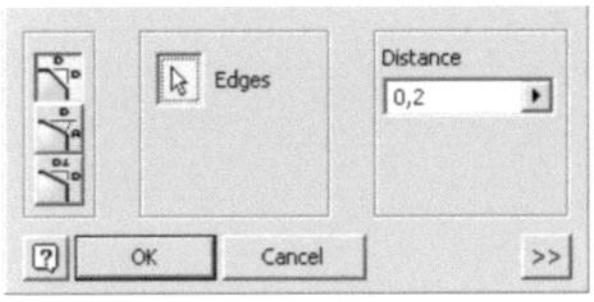

- ***Chamfer***
- Edges: 4 marked cylinder edges
- Distance: 0.2
- OK

6.8.6 Threaded holes for the piston rod fixture

Finally we create the two ***Threads***, which will later be needed to connect the bearing cups.

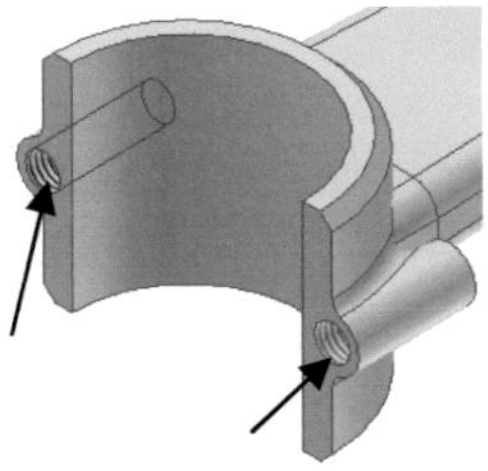

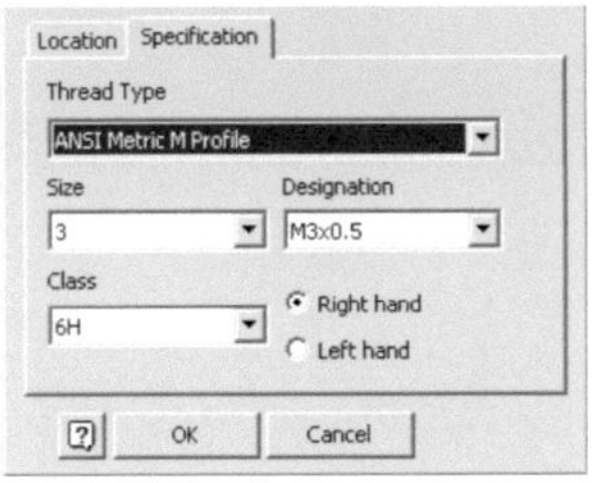

- ***Thread***
- Area: Marked holes
- Thread Type: ISO Metric profile
- Size: 3 - Right hand
- Designation: M3 x 0.5
- Class: 6H
- OK

Note: If you are using ***inch*** units, please transform the metric ones to inch system.

> **Tip**: Should sketches or planes still be visible after finishing the part you can hide them in the model tree with **RMC** > **Visibility**. You should generally check on that when a part is finished so you don't lose track later in the assembly.

topside-piston-rod.ipt now is complete. ▪ **Save** and ✖ **Close** the file.

6.9 Part CRANKSHAFT HOLDER
6.9.1 Basics of the part

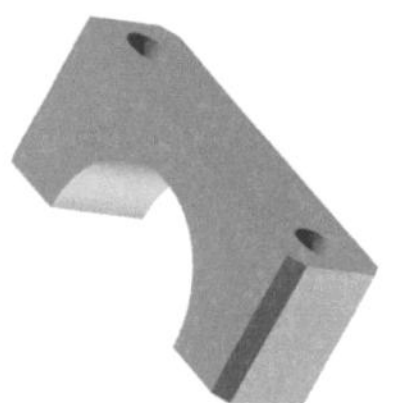

The crankshaft mount is the lower mounting cup for our crankshaft.

It is screwed to the motor block and there it secures the crankshaft against radial and axial drifts.

6.9.2 Volume shape by Extrusion

Create a ▭ **New Part** and ▪ **Save** it as **crankshaft-mount.ipt** in your project folder. Project the X-Y-Z-axes and draw the shown basic outline.

Then use the command ▭ **Extrude** to create the volume shape.

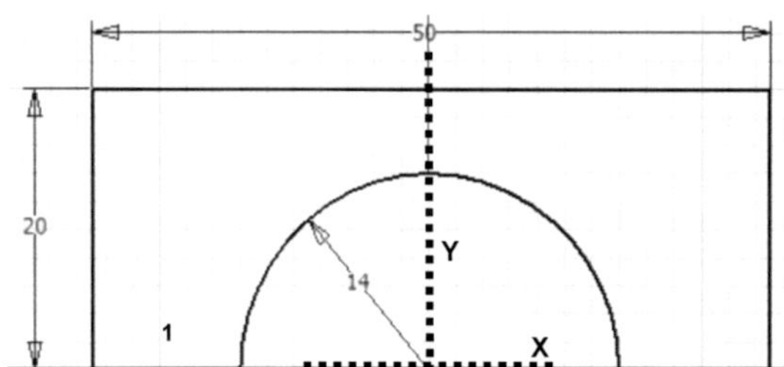

> ➢ ▭ **Project Geometry** (X-Y-Z-axes)
> ➢ Draw shown outline
> ➢ ✔ **Finish Sketch**

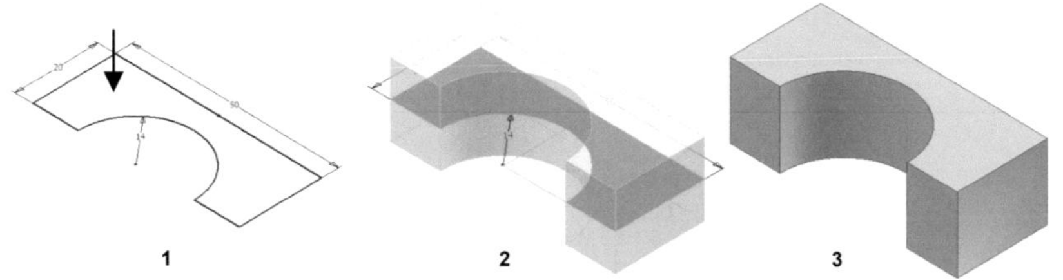

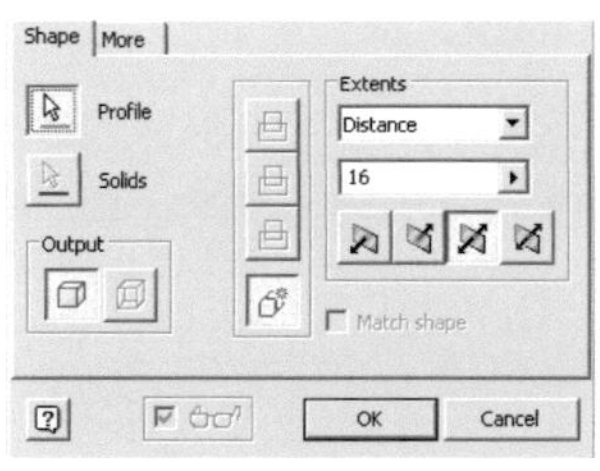

> 🗐 **Extrude**
> Profile: Area (picture 1)
> Operation: Join
> Extents: Distance 16
> Direction: Symmetrical
> [OK]

6.9.3 Through-holes for the screw connection

To later be able to tightly secure the crankshaft holder to the motor housing we need two through-holes. Create a new 🖊 **2D Sketch** on the marked area (picture 1) and place two ✛ **Points**. Then create the two 🗐 **Through holes**.

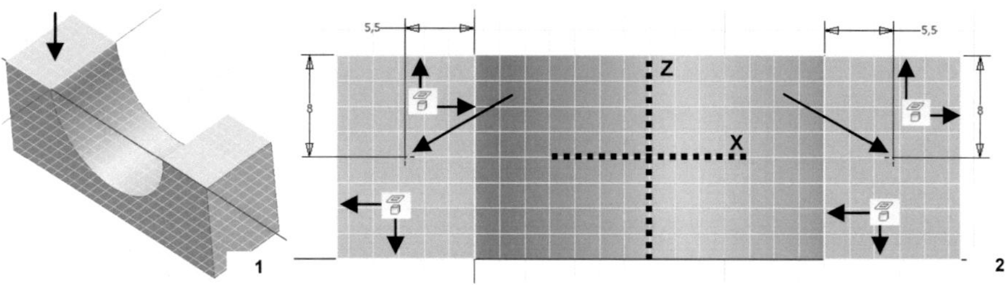

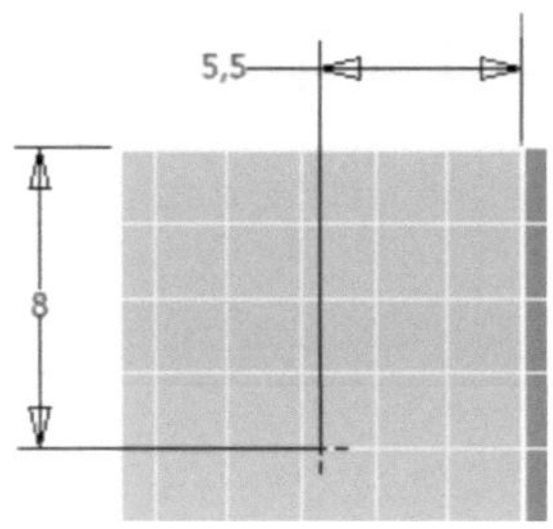

> 🖊 **2D Sketch** (on marked area picture 1)
> 🗗 **Project Geometry** (marked edges picture 2)
> Place two ✛ **Points** according to sketch
> ✔ **Finish Sketch**

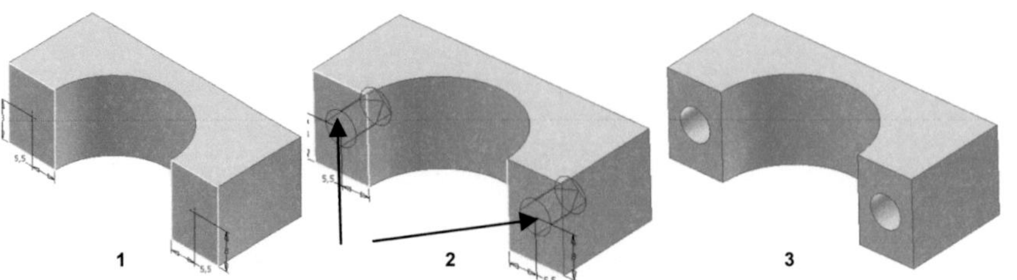

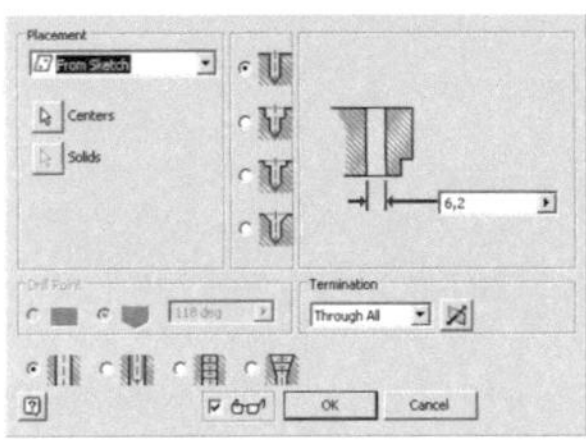

> ⊙ **Hole**
> ➢ Placement: From Sketch
> ➢ Termination: Through All
> ➢ Diameter: 6.2
> ➢ OK

6.9.4 Chamfer the sides

To finish we chamfer the marked 4 sides.

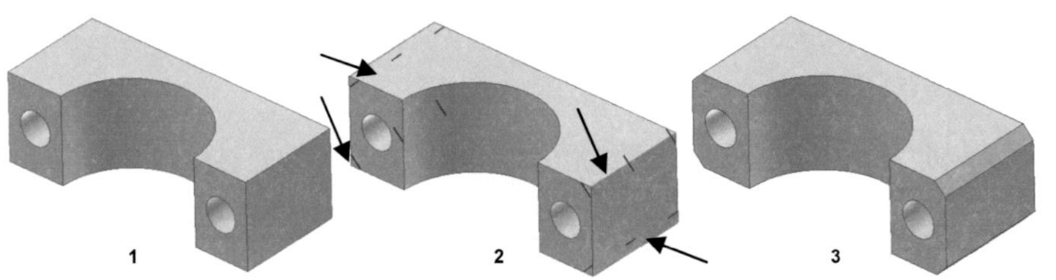

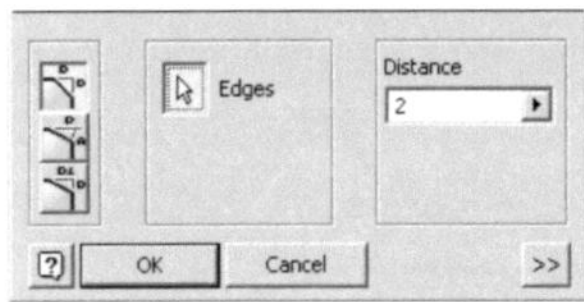

> ➢ ⬙ **Chamfer**
> ➢ Edges: 4 marked edges
> ➢ Distance: 2
> ➢ OK

crankshaft-mount.ipt now is complete. 💾 **Save** and ✖ **Close** the file.

6.10 Part MOTOR HOUSING
6.10.1   Basics of the part

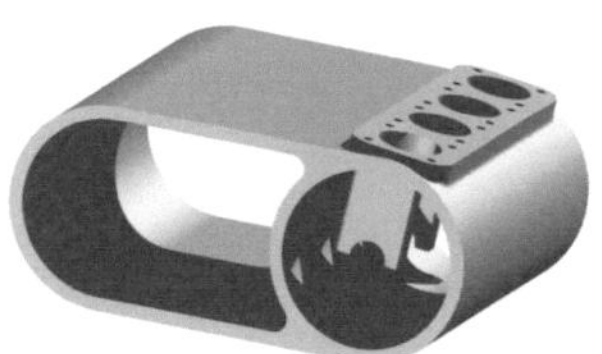

The motor housing (crankcase housing) stores the crankshaft, mounts the transmission and guides the pistons.

It will consist of three main areas (transmission mount, crankshaft area and connector flange for the cylinder block).

6.10.2   Volume shape by Extrusion

For the basic body we create a ⬚ **New Part** and 🖫 **Save** it as **motor-housing.ipt** in the project folder.

Draw the shown basic outline and then use the command ⬚ **Extrude** to create the volume shape. Make sure you have symmetrical extrusion of the volume shape.

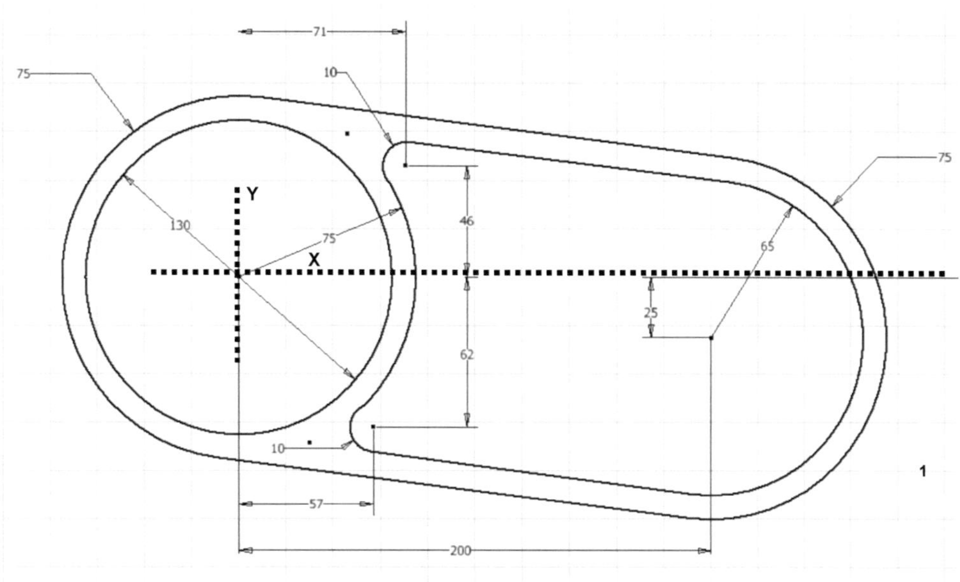

> ⬚ **Project Geometry** (X-Y-Z-axes)
> Draw shown outline
> ✔ **Finish Sketch**

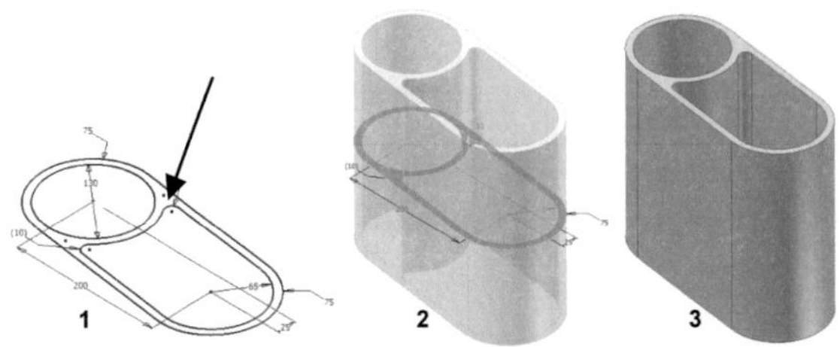

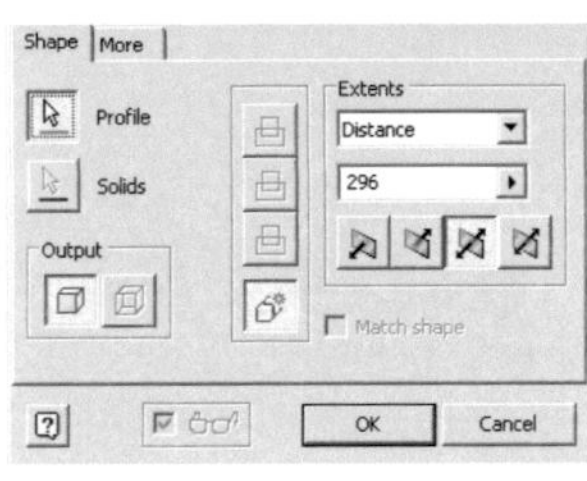

> 🗎 **Extrude**
> Profile: Outline
> Operation: Join
> Extents: Distance 296
> Direction: Symmetrical
> [OK]

6.10.3   Bearings for the crank shaft

The crankshaft bearing will consist of 5 bearing elements. For the first element create a new 🖉 **2D Sketch** on the marked side area and then draw the shown outline.

First project the X-Y-Z-axes and the marked edge (picture 2). Then create the volume shape by using the command 🗎 **Extrude**.

> 🖉 **2D Sketch** (on shown area picture 1)
> 🗗 **Project Geometry** (X-Y-Z-axes and marked edge picture 2)
> Draw shown outline
> ✔ **Finish Sketch**

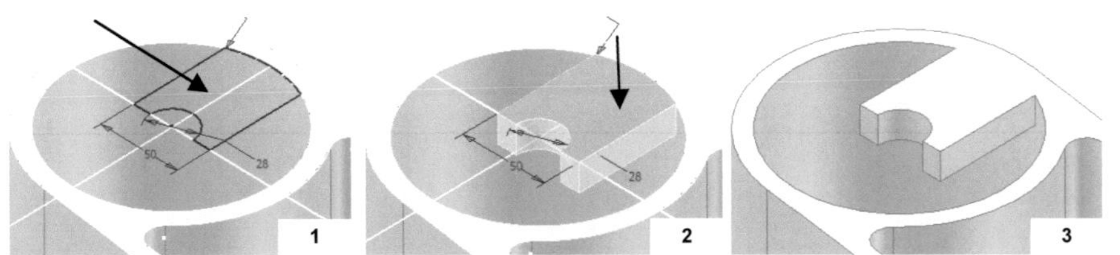

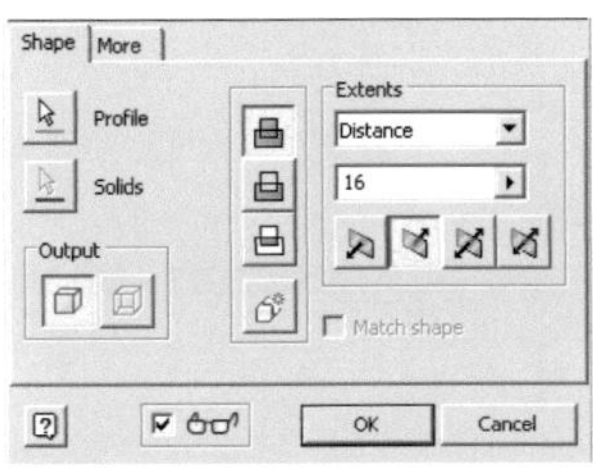

> 🗋 **Extrude**
> Profile: Outline
> Operation: Join
> Extents: Distance 16
> Direction: Shown (picture 2)
> [OK]

6.10.4 Threaded holes for screw joints

For a connection with the crankshaft mount we now need two threaded holes. On the marked area (picture 2) create a new ✍ **2D Sketch**, project the marked edges in picture 1 and place the two hole points.

Then create the two threaded holes by using the command ⬚ **Hole**.

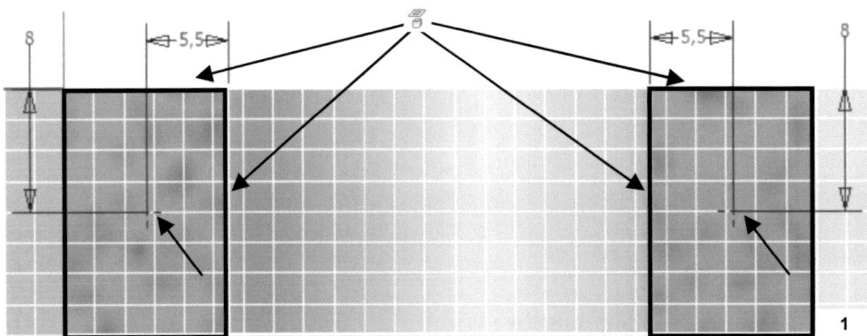

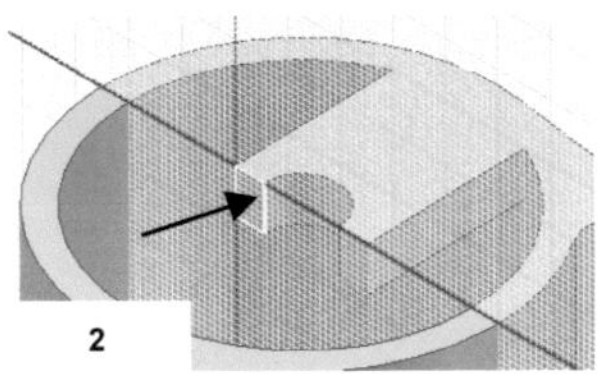

> ✍ **2D Sketch** (on marked area picture 2)
> Slice the sketch with **F7**
> ⬚ **Project Geometry** (marked edges picture 1)
> ⊹ **Point** (place 2 points as shown)
> ✓ **Finish Sketch**

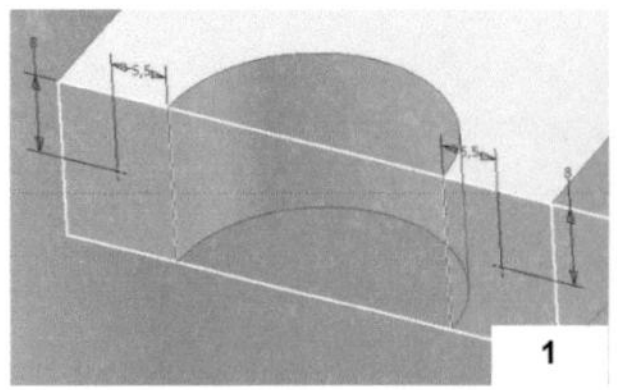 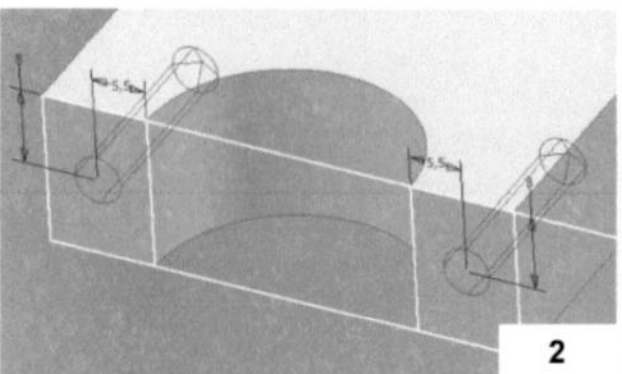 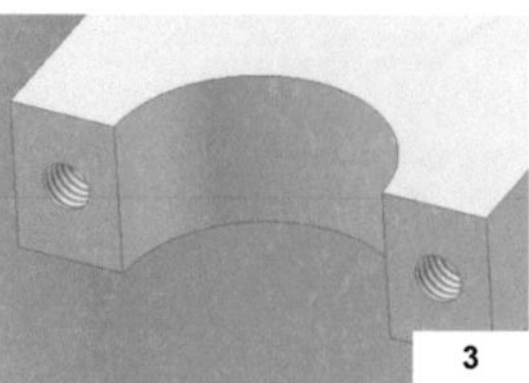

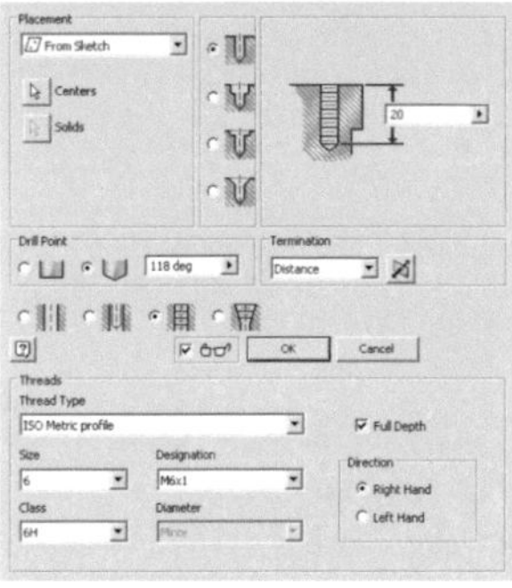

> **Hole** (with thread)
> Placement: From Sketch
> Termination: Distance 20
> Thread Type: ISO Metric profile
> Size: 6 - Right hand
> Designation: M6 x 1
> Class: 6H
> Depth: Full
> OK

Note: If you are using **inch** units, please transform the metric ones to inch system.

6.10.5 Chamfers on the crankshaft bearing

Chamfer the 4 marked edges.

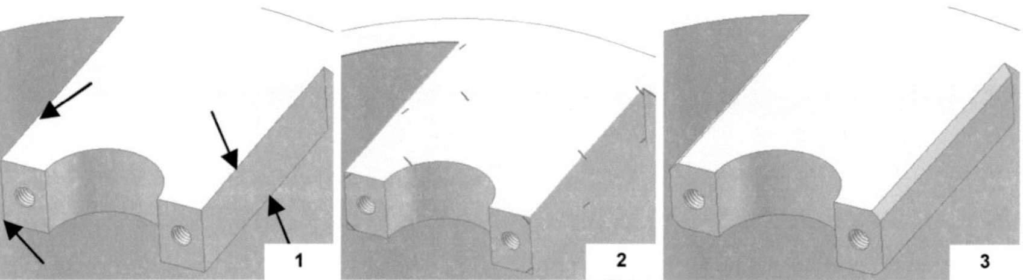

> **Chamfer**
> Edges: 4 marked edges (picture 1)
> Distance: 2
> OK

6.10.6 Arranging the crankshaft bearing

The first bearing element is created. Now it is time to create the other 4 crankshaft bearing elements. For this you use the command **Rectangular Pattern**. Make sure that you really choose all marked elements shown in picture 1 (extrusion, holes and chamfers).

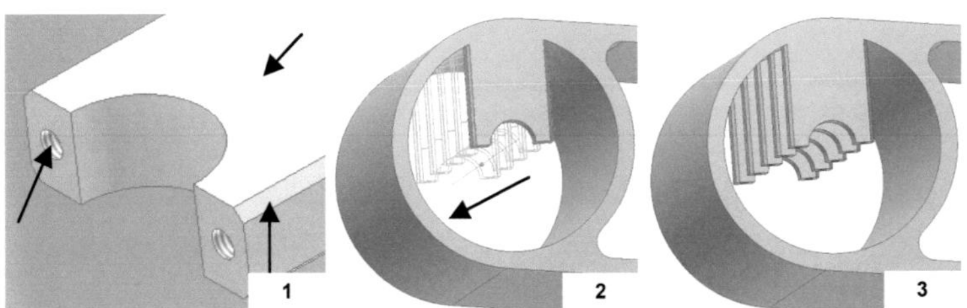

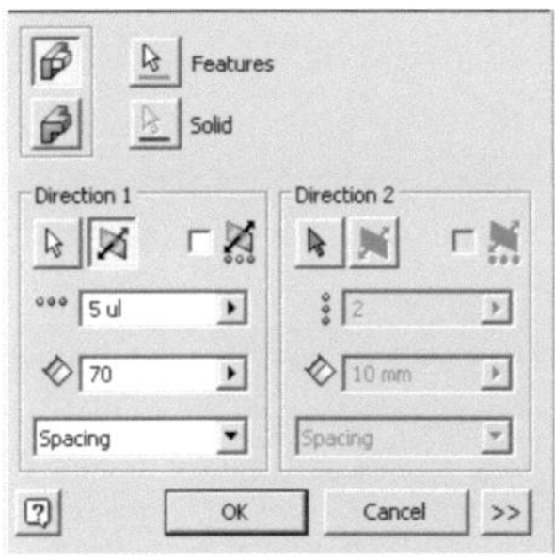

- ⊞ *Rectangular Pattern*
- Elements: Created extrusion, holes and chamfers
- Direction: Y-axis
- Quantity: 5
- Distance: 70
- OK

6.10.7 Seal flanges to the cylinder head

To create the seal flange we need a new plane parallel to the X-Z-plane in the shown direction.

Create a new ✎ *2D Sketch* on it and draw the shown outline (symmetrically to the coordinate's origin). Then create the base body by using the command ⬛ *Extrude*.

Make sure you use the setting *To Next* during extrusion. The volume shape has to exactly close with the existing base body.

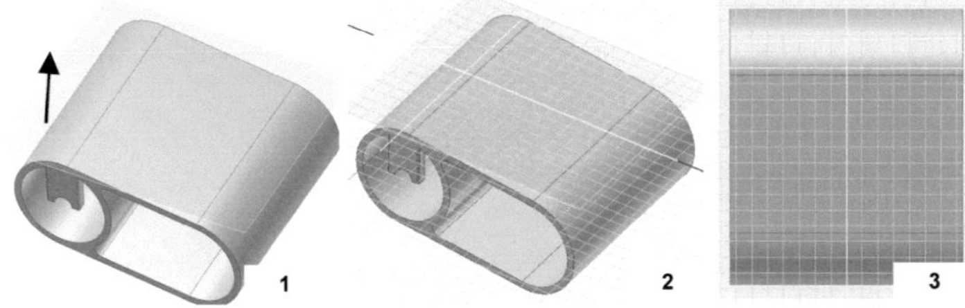

- ⬛ *Plane*
- Click on the X-Z-plane
- Dimension: 80 (to X-Z-plane in showed direction)
- ✓

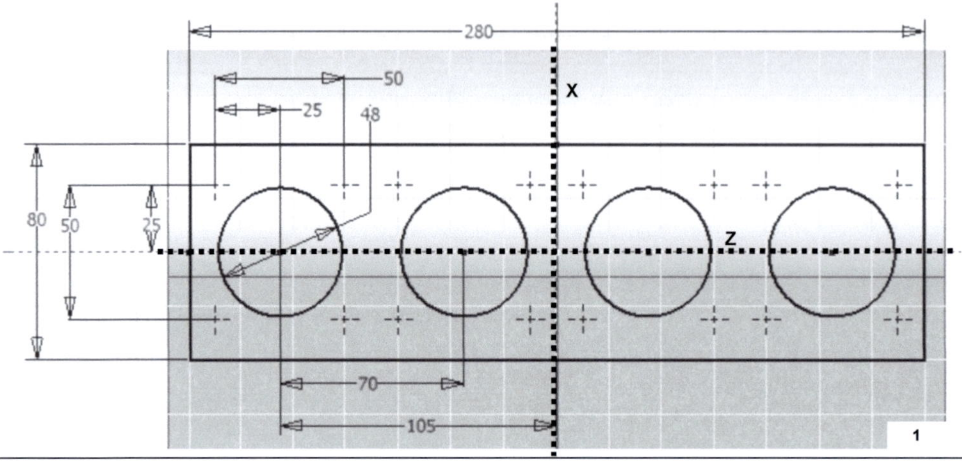

> ✎ **2D Sketch** (on new plane)
> ⛶ **Project Geometry** (X-Y-Z-axes)
> Draw shown geometry (don't forget the 16 ✛ **Points**!)
> ✔ **Finish Sketch**

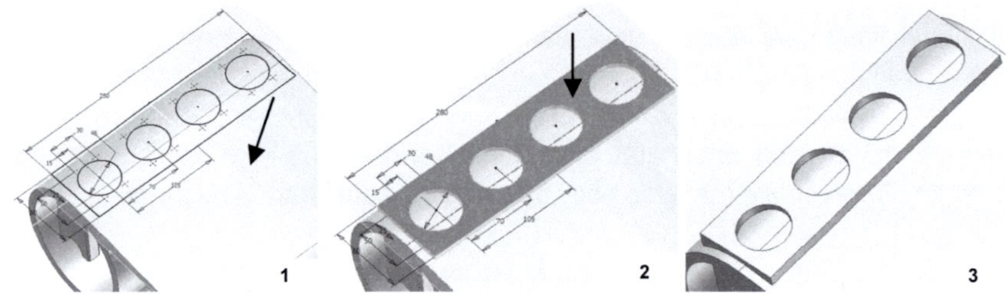

> ⬜ **Extrude**
> Profile: Marked outline (picture 2)
> Operation: Join
> Extents: To Next
> Terminator: Top surface engine case (marked pic. 2)
> ⬜ OK

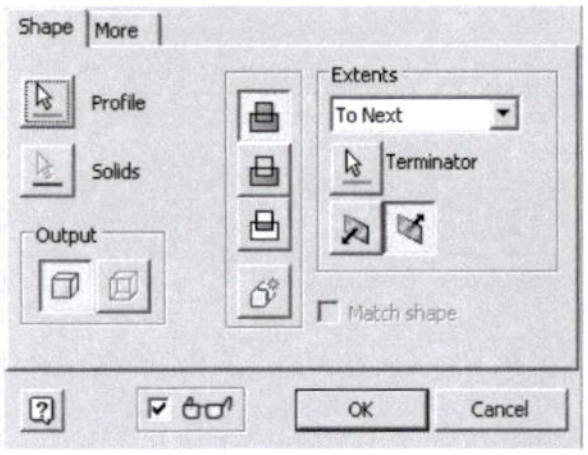

6.10.8   Holes for the Sleeves

For the following two elements (holes for sleeves and threaded holes for the cylinder head screws) we need the last created sketch again (**RMC** > **Share Sketch**). Create a copy body from the 4 marked circles by using the command ⬜ **Extrude**.

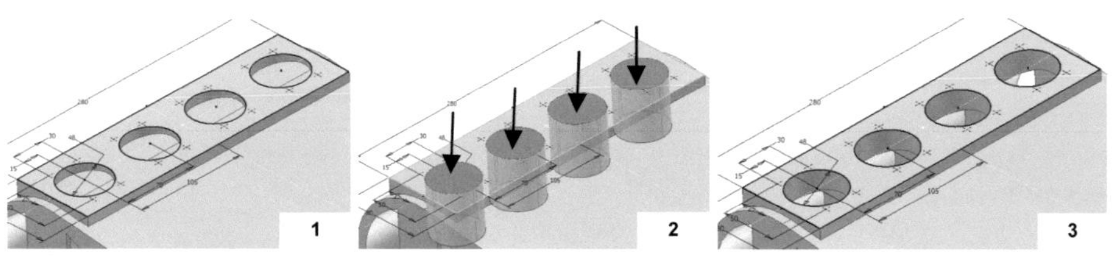

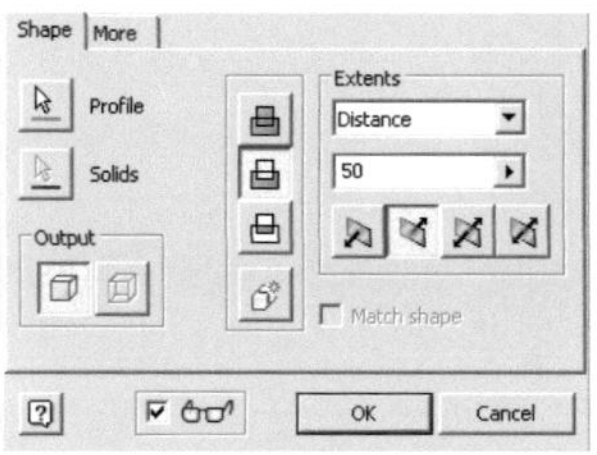

> 📦 **Extrude**
> Profile: Marked 4 circles (picture 2)
> Operation: Cut
> Extents: Distance 50
> Direction: Shown (picture 2)
> ⬛ OK

6.10.9 Threaded holes for the cylinder head screws

Then we create the 16 ⬤ **Threaded holes** for the cylinder head screws.

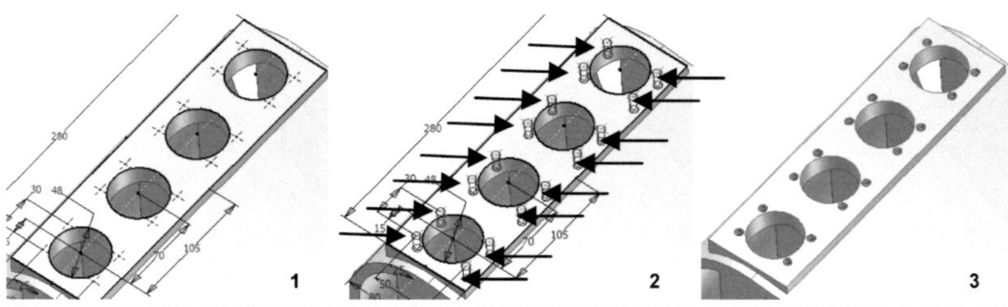

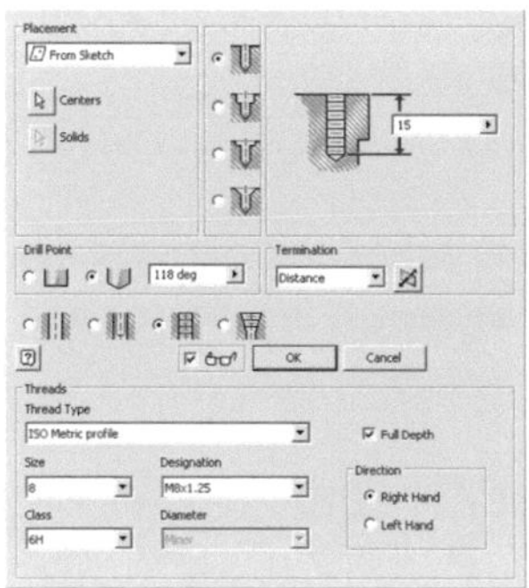

> ⬤ **Hole** (with thread)
> Placement: From Sketch
> Termination: Distance 15
> Thread Type: ISO Metric profile
> Size: 8 - Right hand
> Designation: M8 x 1.25
> Class: 6H
> Depth: Full
> ⬛ OK

Note: If you are using **inch** units, please transform the metric ones to inch system.

As you don't need the visible sketch any longer go into the model tree by using **RMC** and remove the hook under **Visibility**.

6.10.10 Rounding of the transition area

Then we use the command 📄 *Fillet* to design the marked edges and the transition area to the seal flange. Make sure that first the 4 corners (picture 1) are placed, then the revolving rounding (picture 2).

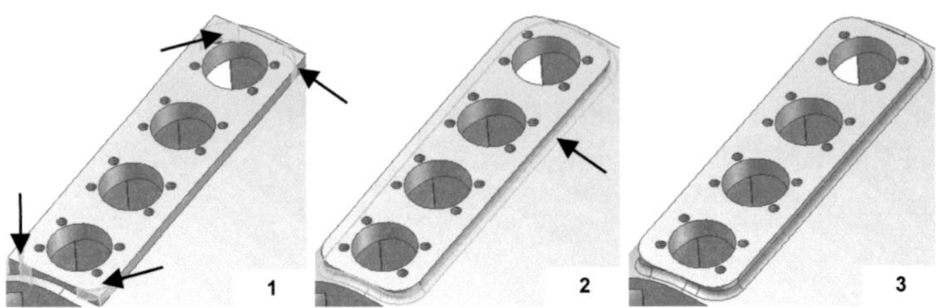

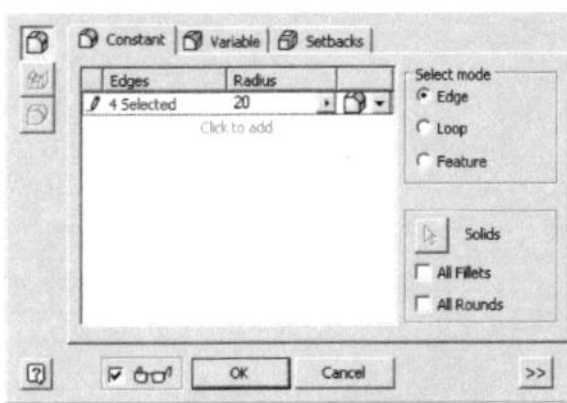

> 📄 *Fillet*
> Edges: 4 marked edges (picture 1)
> Radius: 20
> Cilck to add
> Edges: 1 marked edge (picture 2)
> Radius: 5
> OK

Tip: With the command 📄 *Fillet* the order of the rounding is important in this case. Additional work while choosing the edges might be necessary. By skilful combination of the rounding order work steps can be simplified.

---

motor-housing.ipt now is complete. 💾 *Save* and ❌ *Close* the file.

6.11 Part CYLINDER BLOCK
6.11.1 Basics of the part

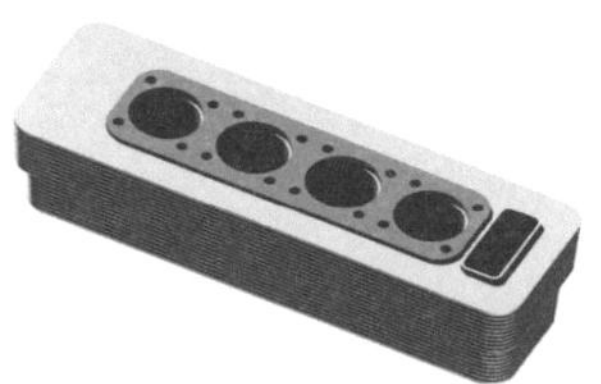

The cylinder block (engine block) serves for cooling of the combustion chambers as well as seat of the sleeves and belt duct.

The cylinder block is located between engine casing and cylinder head and generally is made of cast iron, aluminum or ductile cast iron.

Create a 🗋 *New Part* and 💾 *Save* it as *Cylinder-block.ipt* in the project folder. Project the axes and then draw the shown outline. Make sure that the outline is symmetrically to the Y-axis and placed on the X-axis with the bottom line. Then use the command ▯ *Extrude* to create the volume shape.

6.11.2 Volume shape by Extrude

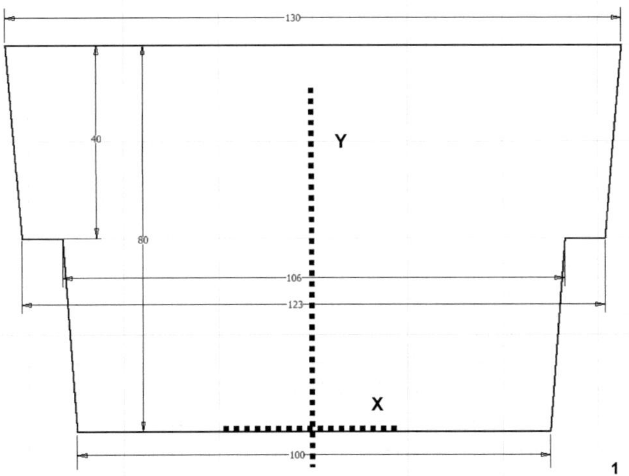

> ⌷ *Project Geometry* (X-Y-Z-axes)
> Draw shown geometry
> ✔ *Finish Sketch*

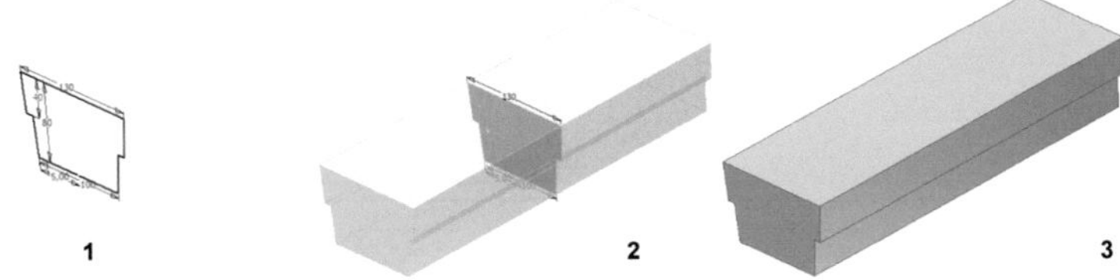

> ▯ *Extrude*
> Profile: Outline
> Operation: Join
> Extents: Distance 400
> Direction: Symmetrical
>   OK

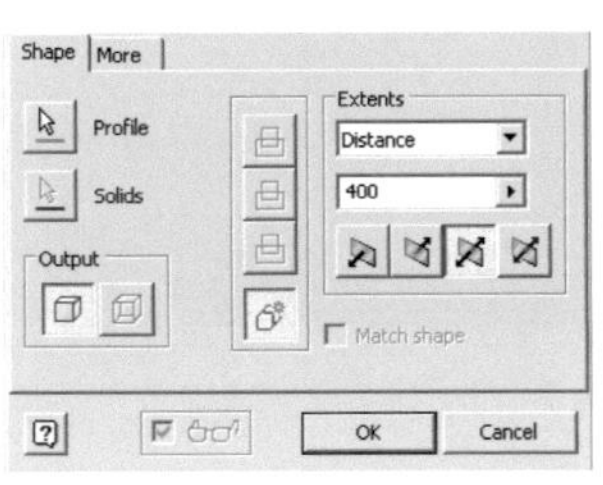

Then create a new ✎ **2D Sketch** on the Y-plane, slice it with **F7** and project the X-Y-Z-axes. Draw the shown outline (the right outline can be mirrored).

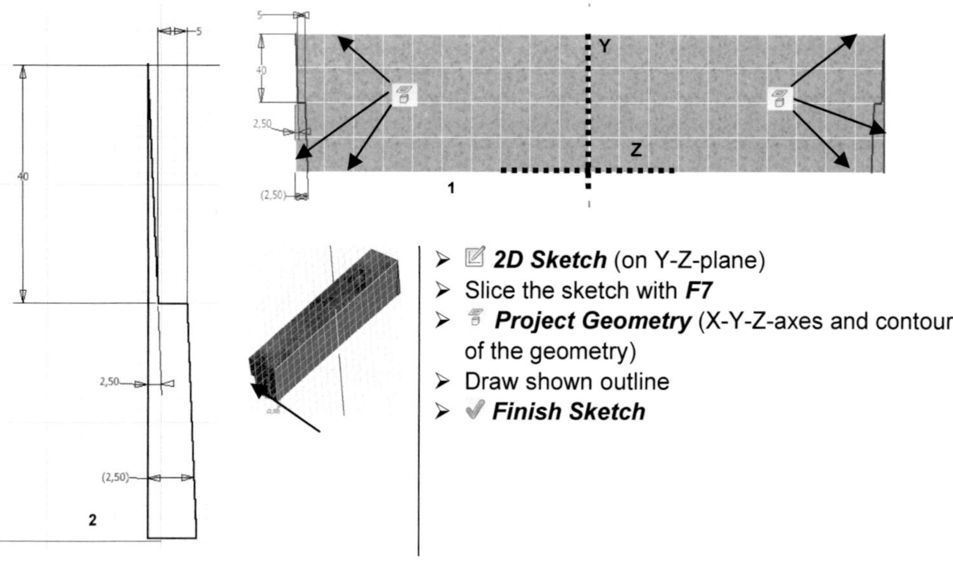

> ✎ **2D Sketch** (on Y-Z-plane)
> Slice the sketch with **F7**
> 🗗 **Project Geometry** (X-Y-Z-axes and contour of the geometry)
> Draw shown outline
> ✔ **Finish Sketch**

Via 🗍 **Extrude** you remove the two geometries from the volume shape.

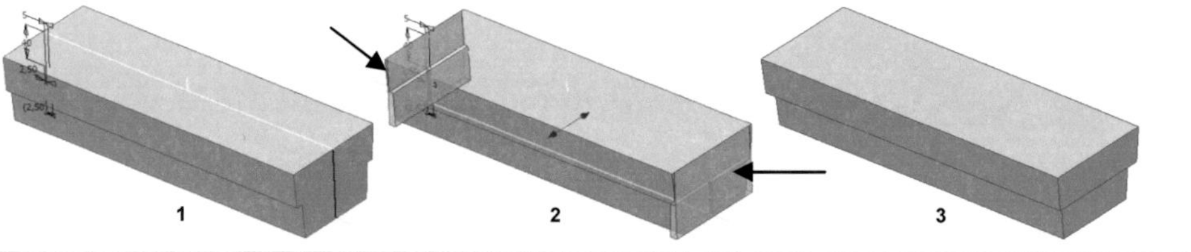

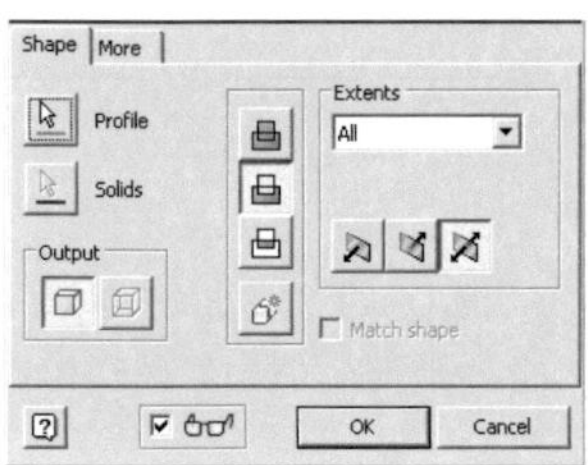

> 🗍 **Extrude**
> Profile: Outlines (both areas)
> Operation: Cut
> Extents: All
> Direction: Symmetrical
> [OK]

Now just rounding the 8 edges.

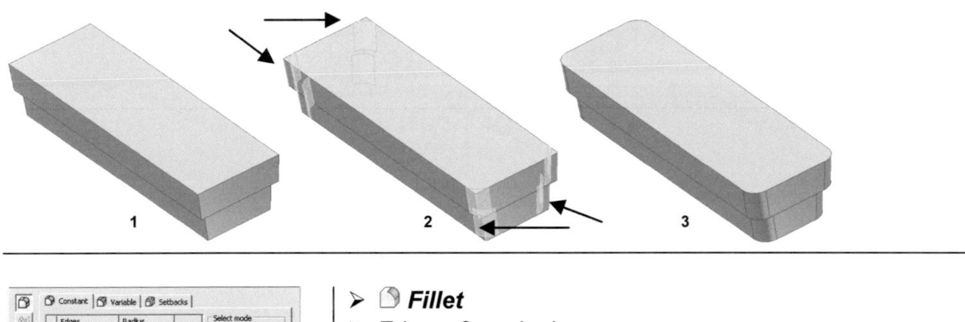

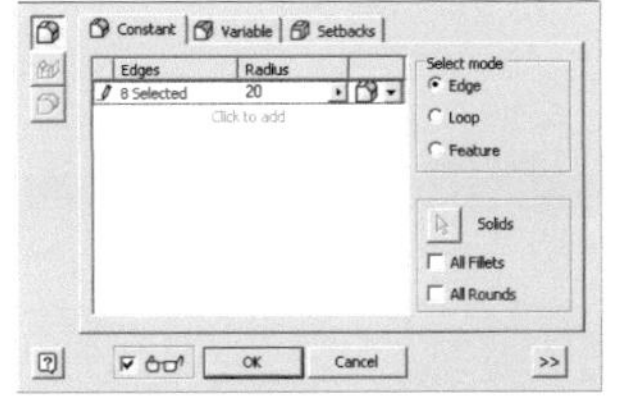

> *Fillet*
> Edges: 8 marked eges
> Radius: 20
> [OK]

6.11.3 Ribs for cooling of the engine

Our sample engine is air cooled. When the base body is created we need the cooling ribs for it. These are created via the command *Sweep*.

The first needed *2D Sketch* we create on the Y-Z-plane. Slice it via *F7* and project the X-Y-Z-axes as well as the marked body edges. Then draw the shown outline. The distance of 8 is in regards to the bottom corner points of the volume shape.

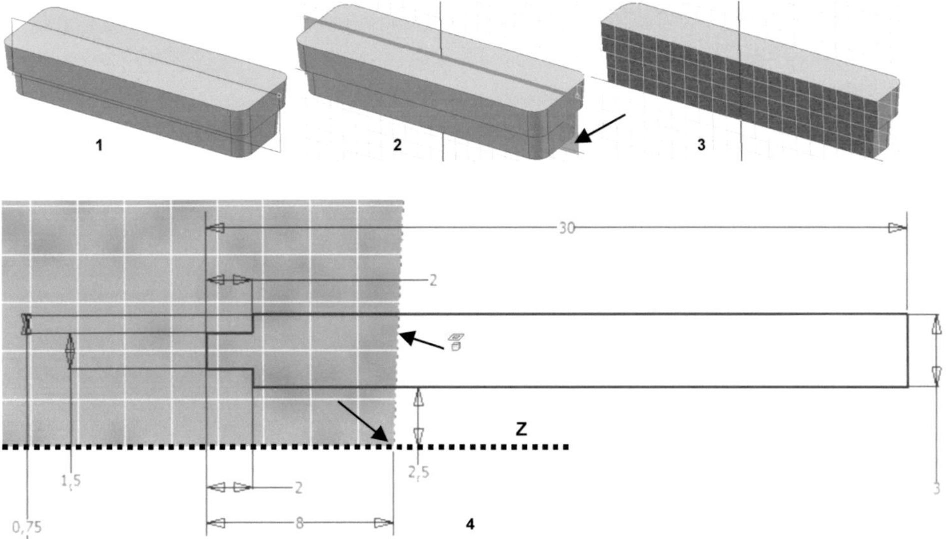

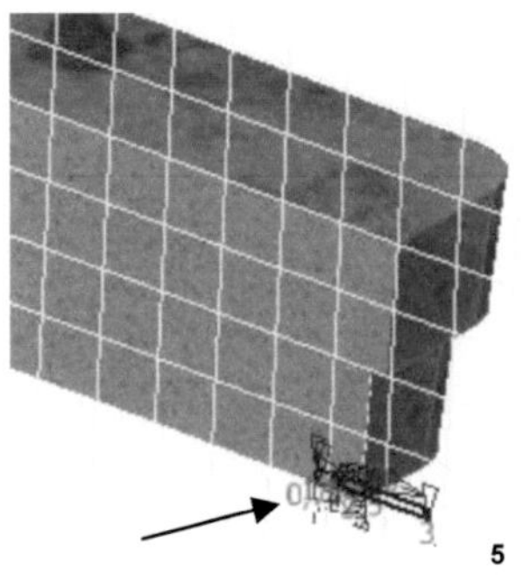

> ➤ 🖉 **2D Sketch** (on marked plane picture 2)
> ➤ Slice the sketch with **F7**
> ➤ 🖫 **Project Geometry** (X-Y-Z-axes and the marked outer edge in picture 4)
> ➤ Draw shown outline
> ➤ ✔ **Finish Sketch**

Tip: When you have the sliced area on screen, the sketch is located on the bottom right side on the outer contour of the existing object (picture 5).

Now we need a second sketch to determine the path for our command 🕃 **Sweep**.

Create a **2D Sketch** on the marked area (picture 1), project the marked edges and finish the sketch. Then sweep the object.

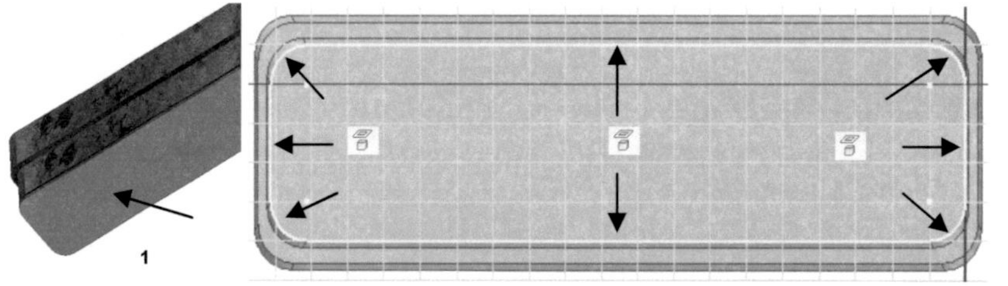

> ➤ 🖉 **2D Sketch** (on marked area picture 1)
> ➤ 🖫 **Project Geometry** (marked 8 edges picture 2)
> ➤ ✔ **Finish Sketch**

Tip: You can mark all edges separately (picture 2) or click once on the area marked in picture 1. Then all adjoining lines are automatically marked.

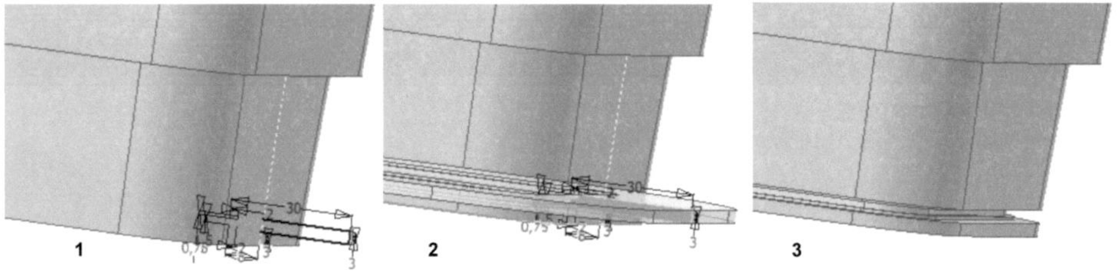

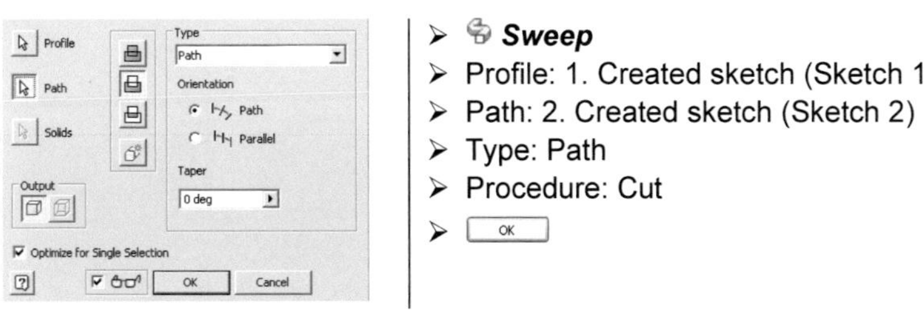

> ⊕ **Sweep**
> Profile: 1. Created sketch (Sketch 1)
> Path: 2. Created sketch (Sketch 2)
> Type: Path
> Procedure: Cut
> [OK]

To create the rest of the cooling ribs we use the command ⠿ **Rectangular Pattern**.

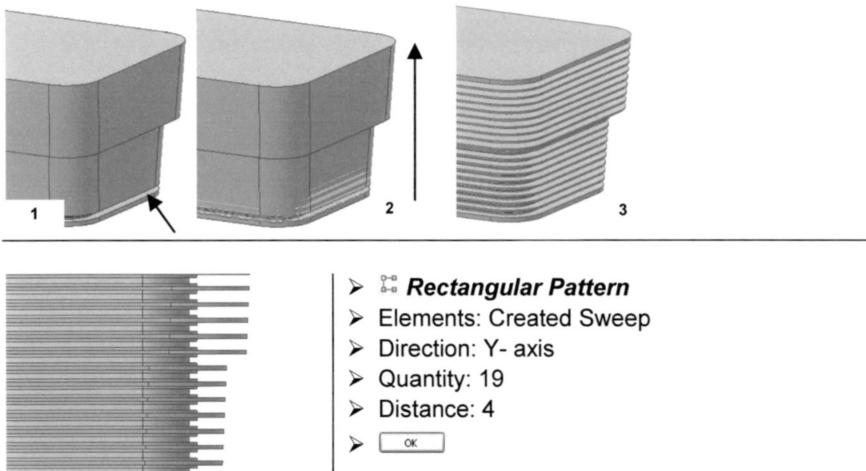

> ⠿ **Rectangular Pattern**
> Elements: Created Sweep
> Direction: Y- axis
> Quantity: 19
> Distance: 4
> [OK]

6.11.4 Holes for sleeves, belts and cylinder head screws

It is time to draw a sketch for the through-holes (sleeves, belts and cylinder head screws) and seal areas.

Create a new ✍ **2D Sketch** on the marked area (topside of the cylinder block) and then draw the shown outline.

Project the X-Y-Z-axes before that and make sure that the 4 cylinder holes are located symmetrically to the X- and Z-axes.

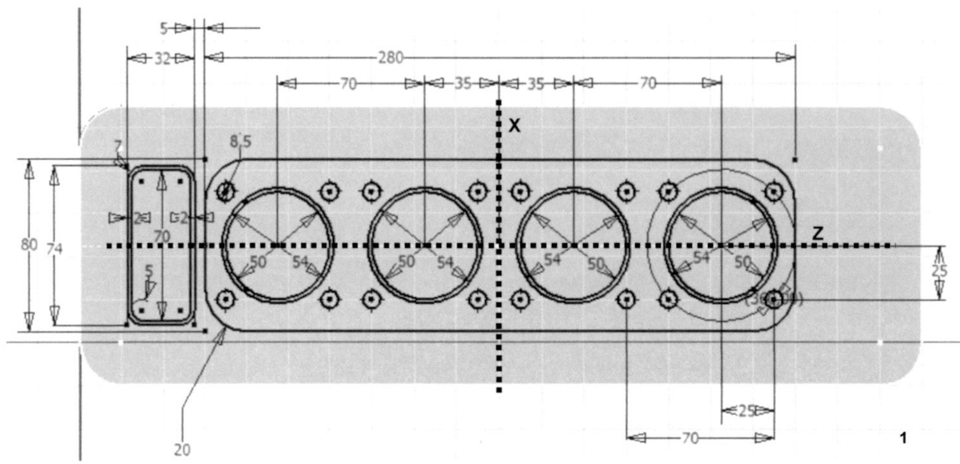

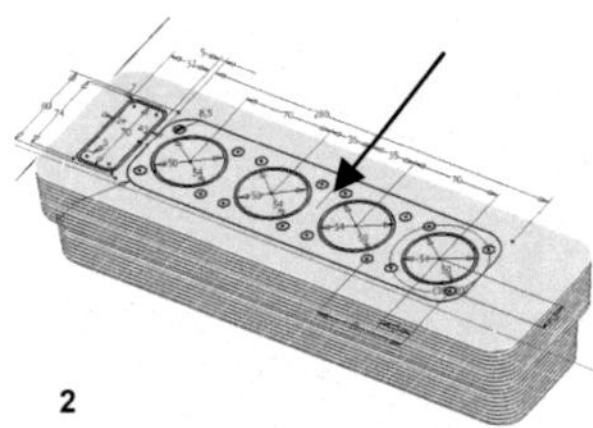

- ➤ 🖉 **2D Sketch** (on marked area in picture 2)
- ➤ 🔗 **Project Geometry** (X-Y-Z-axes)
- ➤ Draw shown outline
- ➤ ✅ **Finish Sketch**

Then the shown notches (4x sleeves, 1x belt and 16x holes for cylinder head screws) are removed by using the command 🗍 **Extrude** (picture 3).

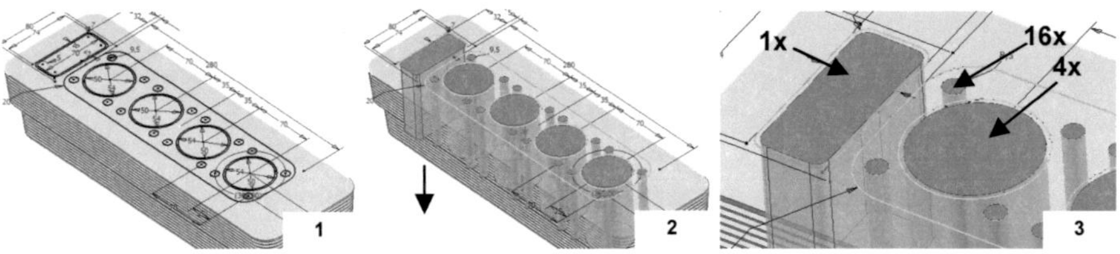

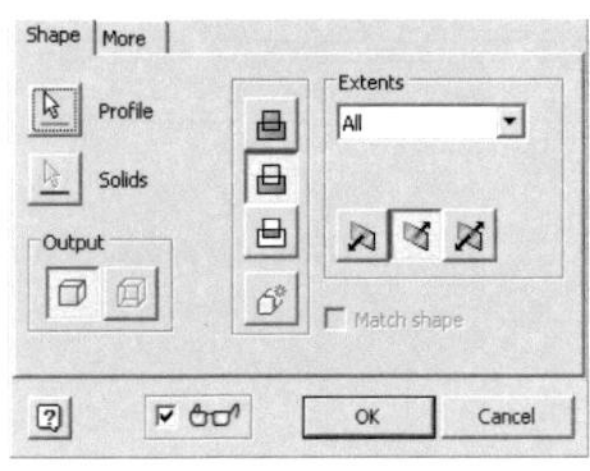

- ➤ 🗍 **Extrude**
- ➤ Profile: Marked areas (picture 3)
- ➤ Operation: Cut
- ➤ Extents: All
- ➤ Direction: Shown (picture 2)
- ➤ [OK]

Tip: For the 4 circles choose only the inner circles (D= 50) for the cylinder bores. The outer circles are used afterwards.

Make the sketch visible (*RMC* > *Visibility*) again and ⬚ *extrude* the 4 marked circles.

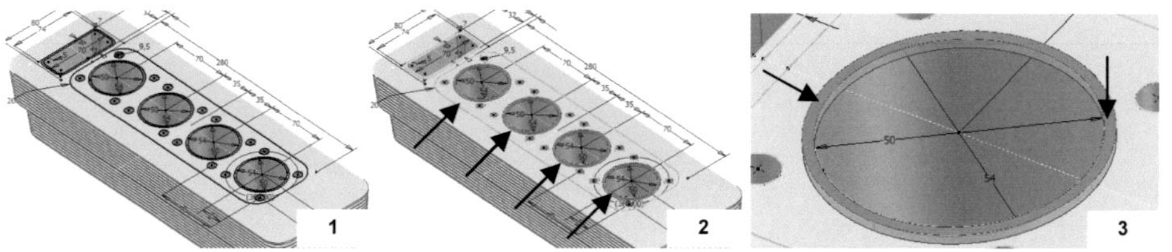

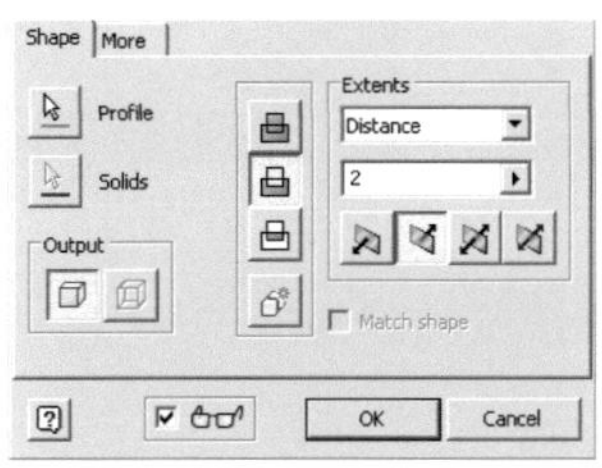

> ⬚ *Extrude*
> Profile: Marked area between D= 50 and D= 54 (pic. 3)
> Operation: Cut
> Extents: Distance 2
> Direction: Shown (picture 3)
> [OK]

6.11.5 Composition surface for the cylinder head

Now we will create the composition surface for the cylinder head and then the composition surface for the belt duct from the existing sketch by using the command ⬚ *Extrude*.

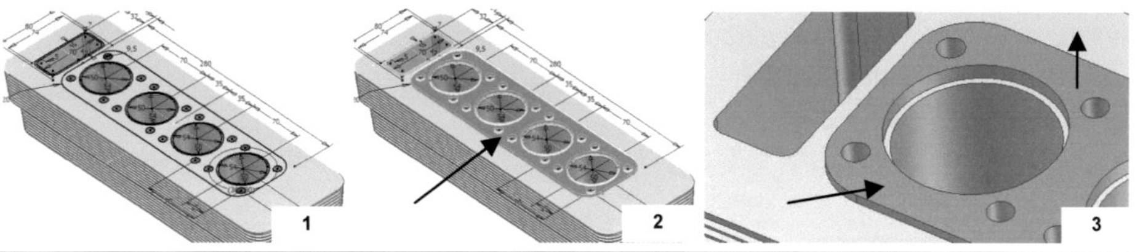

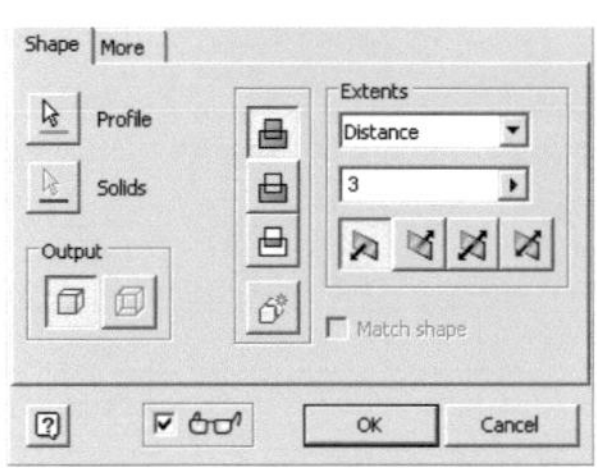

> ⬚ *Extrude*
> Profile: Marked area (picture 2)
> Operation: Join
> Extents: Distance 3
> Direction: Shown (picture 3)
> [OK]

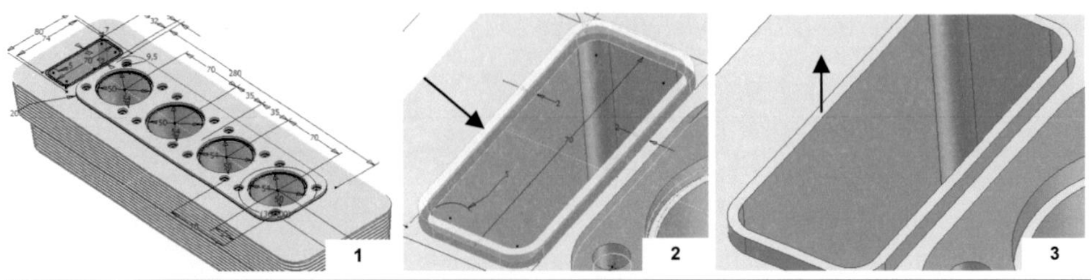

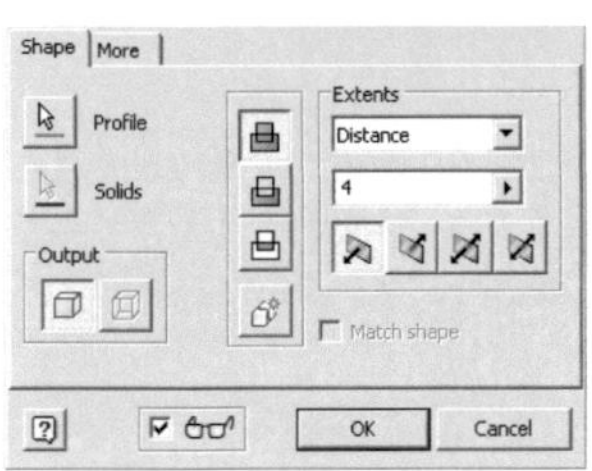

>   *Extrude*
>   Profile: Marked area (picture 2)
>   Operation: Join
>   Extents: Distance 4
>   Direction: Shown (picture 3)
>   OK

6.11.6   Composition surface for the cylinder block

Now the visibility of the sketch can be removed again. The geometry is identical to the geometry of the composition surface to the cylinder head on the opposite side.

Therefore we simply can copy this by using the command   *Project Geometry*. In the view of the drawing area rotate the cylinder block by 180° and create a new   *2D Sketch* on the bottom side of the cylinder block.

Then project the outlines of the two marked seal areas. Make sure not to choose the lines separately, but to click on the respective areas.

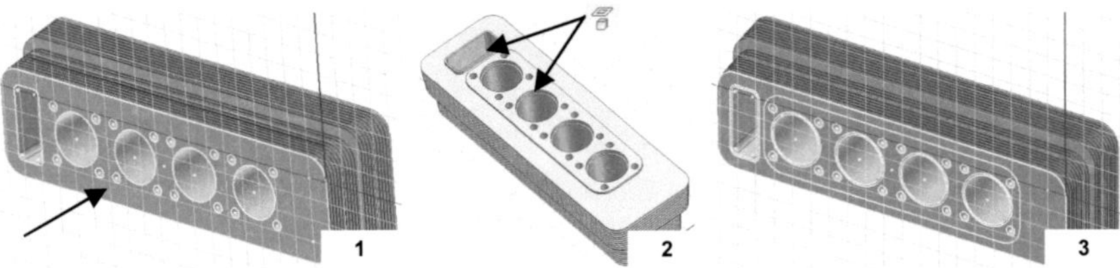

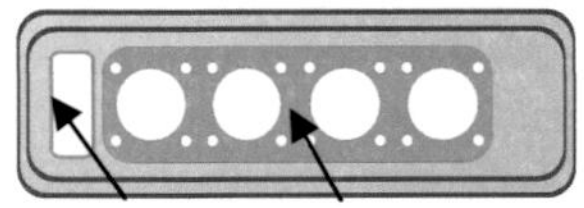

>   *2D Sketch* (on marked area picture 1)
>   *Project Geometry* (outline edges see picture 2)
>   *Finish Sketch*

Then we also create on this side of the cylinder block two ▯ **Extrusions** (one for the belt duct, one for the composition surface).

The sketch has to be made visible again after the first extrusion.

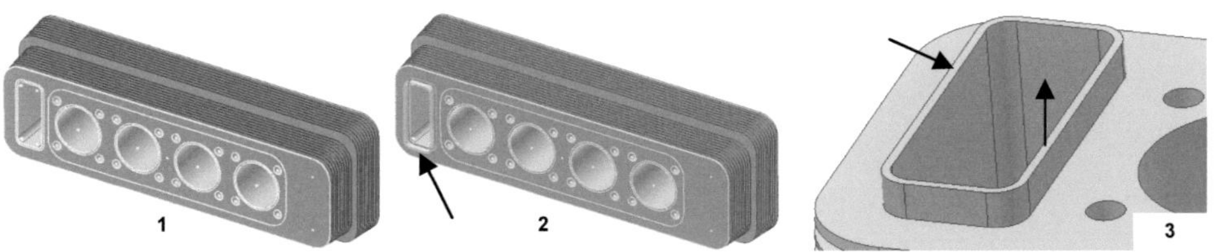

> ▯ **Extrude**
> Profile: Marked area (picture 2)
> Operation: Join
> Extents: Distance 8
> Direction: Shown (picture 3)
> OK

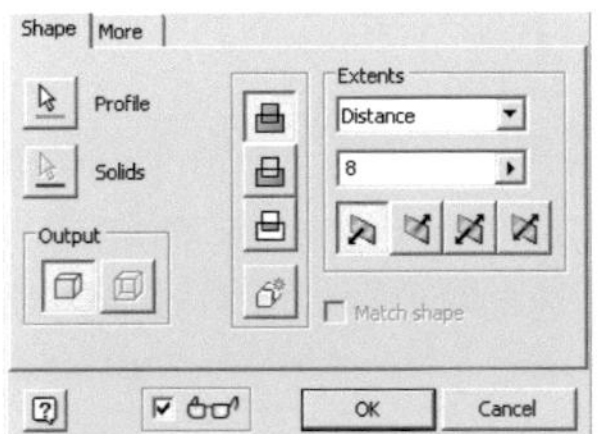

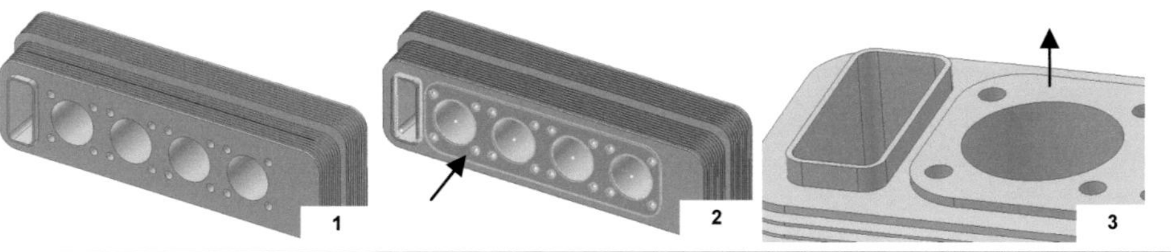

> ▯ **Extrude**
> Profile: Marked area (picture 2)
> Operation: Join
> Extents: Distance 3
> Direction: Shown (picture 3)
> OK

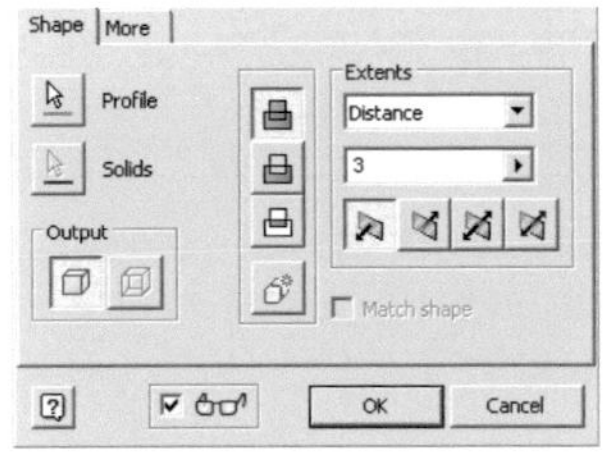

cylinder-block.ipt now is complete. ▯ **Save** and ✕ **Close** the file.

6.12 Part CYLINDER HEAD
6.12.1 Basics of the part

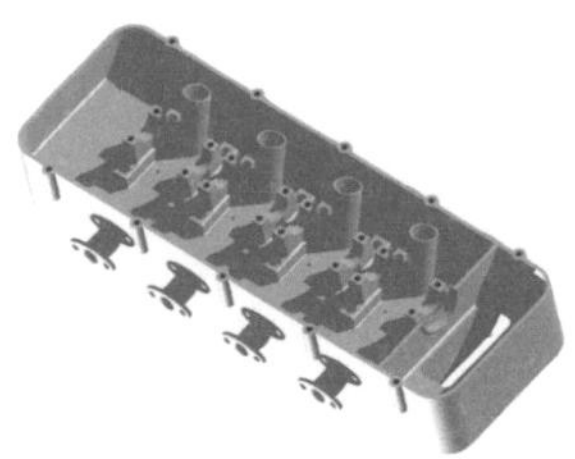

The cylinder head closes the combustion chamber of the cylinders to the top. It stores the camshaft, the valves, the spark plugs, part of the belt duct and forms connection canals between the cylinder spaces and exhaust and intake manifolds.

Generally a cylinder head is made of cast iron, aluminum or ductile cast iron.

6.12.2 Extrapolation from a sketch

Starting from the part **cylinder-block.ipt** we will use the existing outlines (composition surface cylinder head and belt duct) to build our new part on. **Open** the file **cylinder-block.ipt** and create a new **2D Sketch** on the seal area of the cylinder head. Use the command **Project Geometry**, to adopt the two marked composition surfaces (composition surface cylinder head and belt duct).

In the still open sketch mark all just created outlines (draw a window over all lines with **LMC**) and copy these (**Strg+C** or **RMC > Copy**). Finish the sketch but keep the part open.

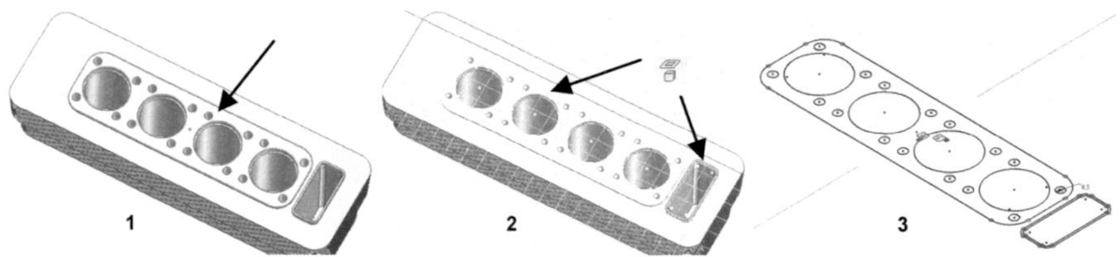

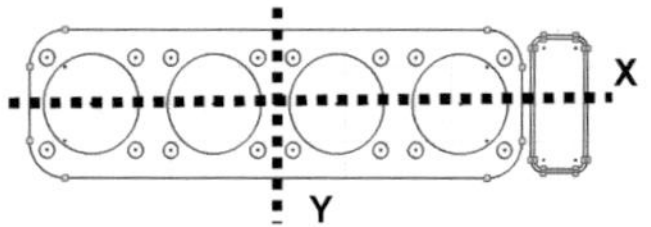

> ➤ **2D Sketch** (on marked area picture 1)
> ➤ **Project Geometry** (outside edges of the geometry)
> ➤ **RMC > Copy** (newly created sketch)
> ➤ **Finish Sketch**
>
> ➤ Keep part **cylinder-block.ipt** still open
>
> ➤ **Strg+V** or **RMC > Paste**

Create a **New Part** and **Save** it as **cylinder-head.ipt** in the project folder. Integrate the just copied outlines into the part in the already opened sketch (**Strg+V** or **RMC > Paste**).

Tip: To transfer the geometries of the two composition surfaces to our new part we also could have copied the already existing and several times used basic sketch. As we, however, want to drift our geometries to zero (for clean working) the geometrical dependencies of the basic sketch would just distract us. Therefore we created a new sketch with derivated outlines.

Then we rotate and move the outline within the sketch in a way that the four cylinder holes are located symmetrically to the coordinate's origin.

Use the commands ○ **Rotate** and ✛ **Move**, to get to the desired position. Project the X-Y-Z-axes prior to that.

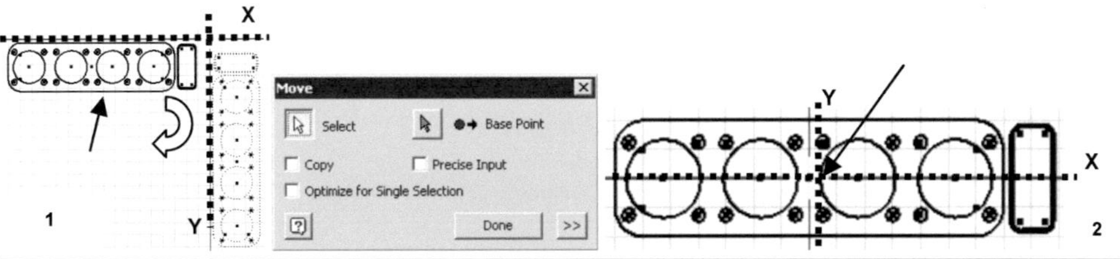

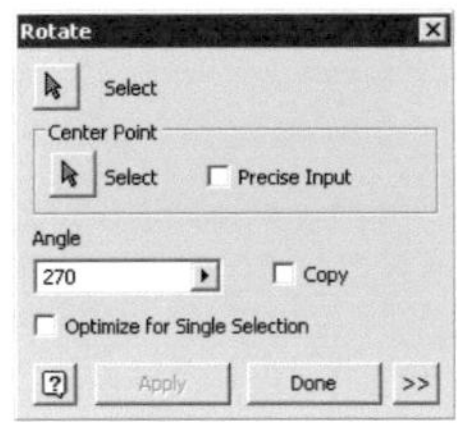

> ○ **Rotate**
> Select: Mark imported outline
> Center Point: Any point
> Angle: 270° (picture 1)
> Apply

> ✛ **Move**
> Select: Mark imported outline
> Base point: the center point of the geometry
> Move to origin of the coordinates (picture 3)
> Apply

> ✔ **Finish Sketch**

Tip: The rotation and drift of the outline shown is just an example. The location of the outline in your sketch, of course, can be different.

You can change back to the part **cylinderblock.ipt** and ✖ **Close** the file <u>without</u> saving. It now is time to create two volume shapes by using the command ▱ **Extrude**.

Start with the seal area to the cylinder block, then the seal area for the belt duct. The sketch has to be reused between both steps (**RMC** > **Share Sketch**).

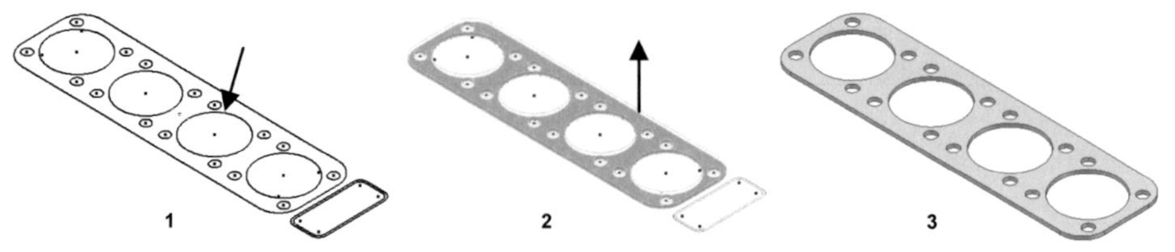

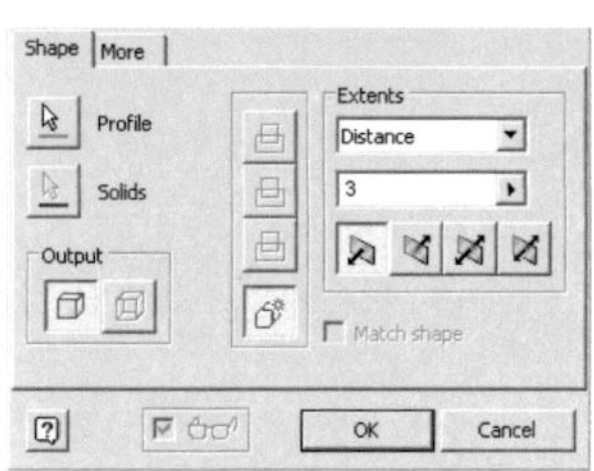

> 🗍 **Extrude**
> Profile: Marked area (picture 1)
> Operation: Join
> Extents: Distance 3
> Direction: Shown (picture 2)
> OK

> **RMC** > **Share Sketch**

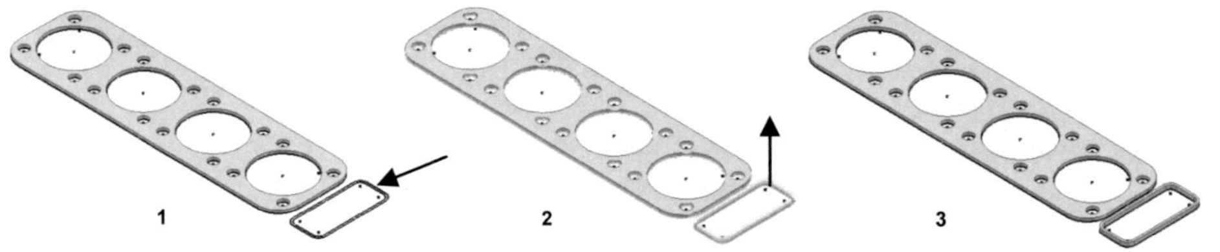

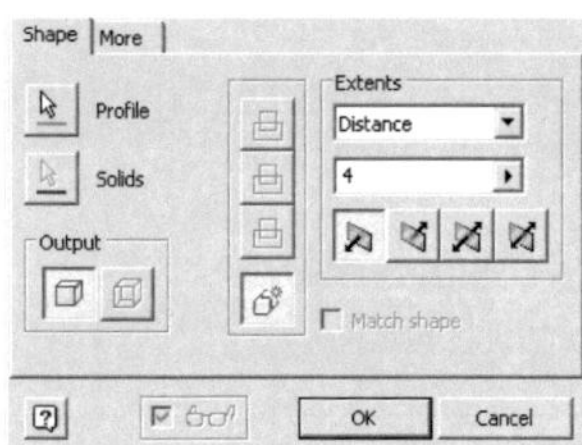

> 🗍 **Extrude**
> Profile: Marked area (picture 1)
> Operation: Join
> Extents: Distance 4
> Direction: Shown (picture 2)
> OK

You can remove visibility of the sketch again. Now rotate the created elements (180°) and create a new 🖉 **2D Sketch** on the backside of one of the two elements. Project the X-Y-Z-axes and draw the shown outline.

Tip: Make sure you use the correct side for the creation of the new sketch. When you look at the new sketch both created elements have to point back. Also make sure of that with the next element, which has to be extruded in the opposite direction.

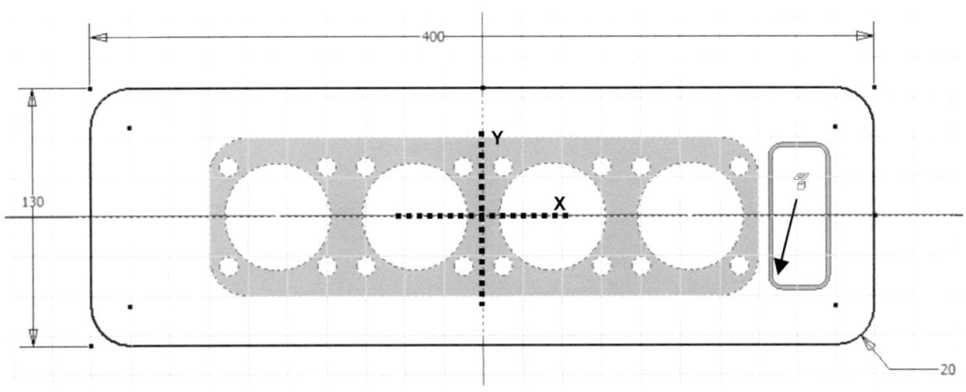

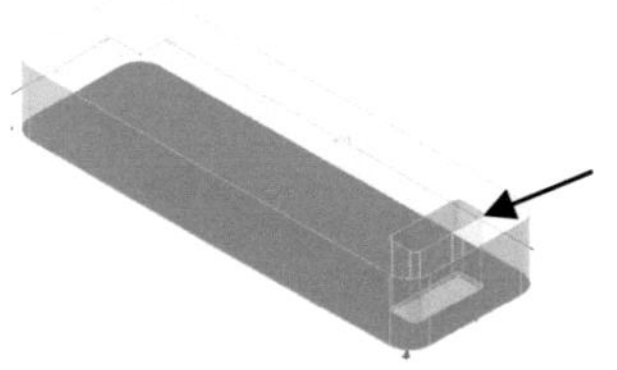

> ✐ **2D Sketch** (on marked area)
> ⌸ **Project Geometry** (X-Y-Z-axes and inner outline edge of the belt duct-seal area)
> Draw shown outline
> ✔ **Finish Sketch**

Tip: It is important to project the inner edges of the belt duct, not the outer ones!

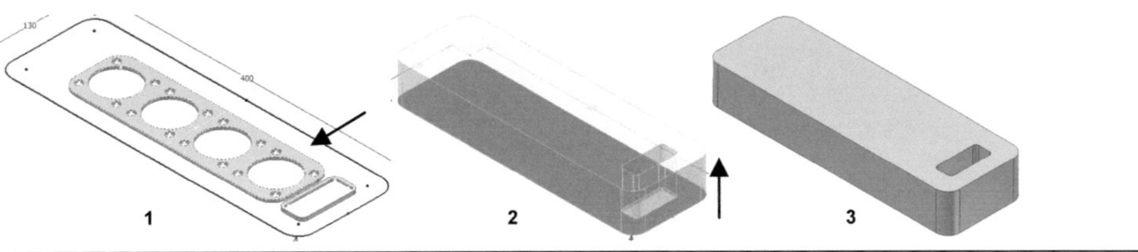

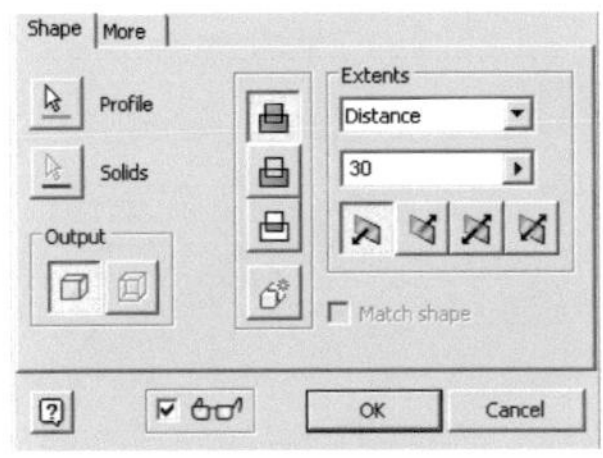

> ▯ **Extrude**
> Profile: Marked area (picture 1)
> Operation: Join
> Extents: Distance 30
> Direction: Shown (picture 2)
> [OK]

6.12.3 Inner and outer walls of the cylinder head

The next step is the creation of the seal walls. Create a new ✐ **2D Sketch** on the marked area. Use the command ⌸ **Project Geometry**, to project the X-Y-Z-axes and the outer edges of the existing outline. Draw the following outline and extrude a volume shape from it afterwards.

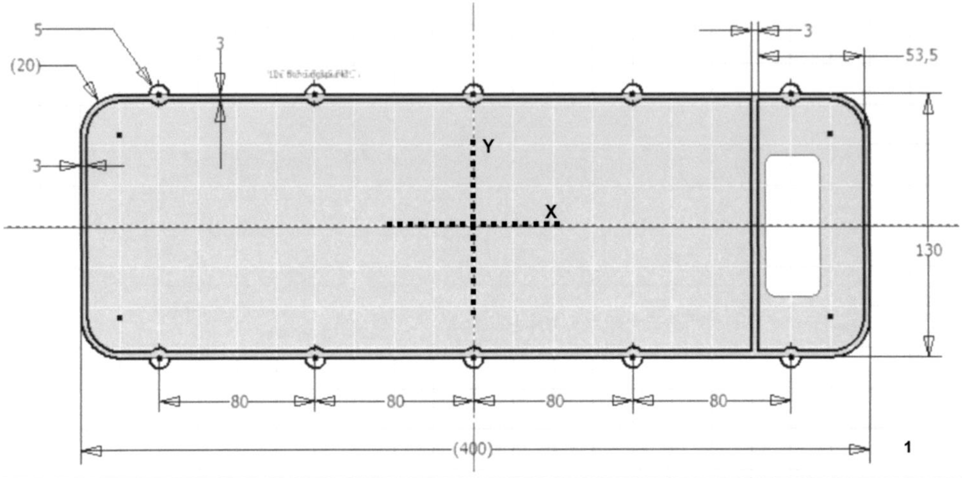

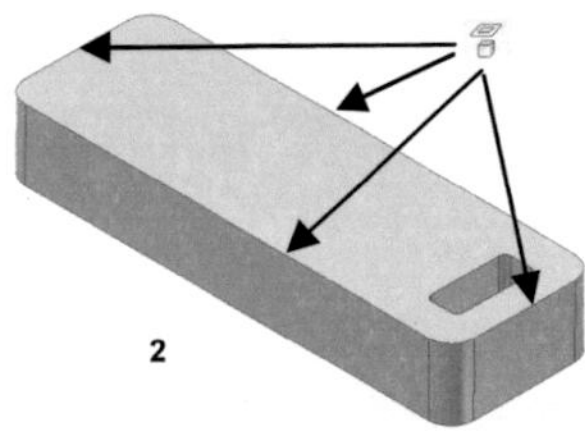

> ☑ **2D Sketch** (on marked area picture 1)
> ☞ **Project Geometry** (X-Y-Z-axes and 4 marked outer edges in picture 2)
> Draw shown outline (picture 1)
> ✔ **Finish Sketch**

Tip: Make sure that the outer lines of the outline to draw closes with the outer edges of the existing volume shape. Use ↗ **Collinear**.

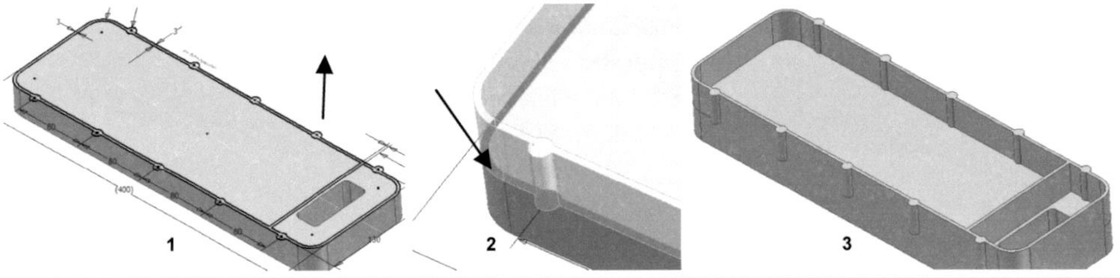

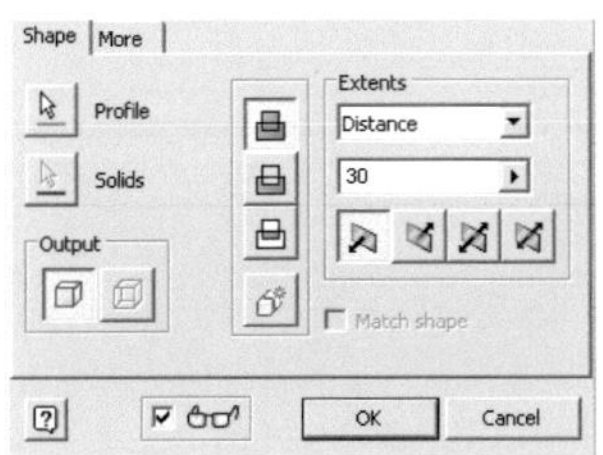

> ⬚ **Extrude**
> ➤ Profile: Marked area (picture 2)
> ➤ Operation: Join
> ➤ Extents: Distance 30
> ➤ Direction: Shown (picture 1)
> ➤   OK

Then rounding the marked edges (5 edges are shown + 5 edges on the other side).

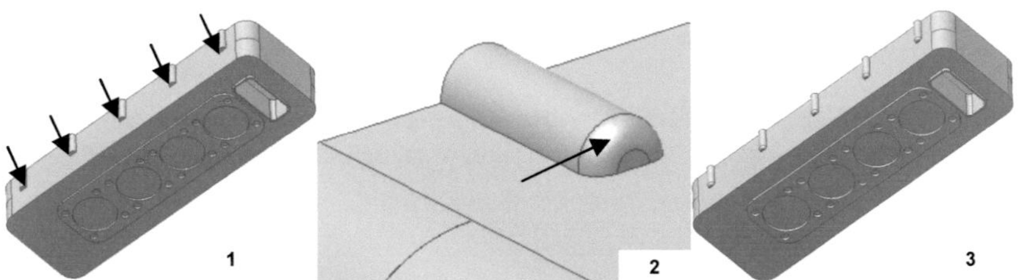

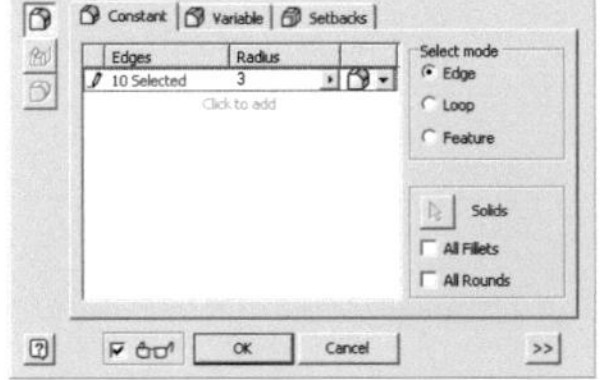

> ➤ ⬚ **Fillet**
> ➤ Edges: 10 marked edges (pictures 1+2)
> ➤ Radius: 3
> ➤   OK

6.12.4 Threaded valve cover holes

For the first threaded hole open the command ⬚ **Hole** and there use the placing type **Concentric**.

Now you first choose the marked composition surface (picture 1), then the marked circle edge (picture 2). The program automatically places the holes concentric to the existing cylinder edge.

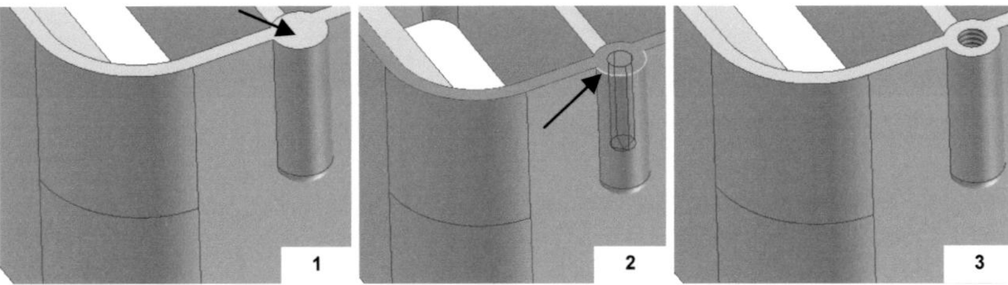

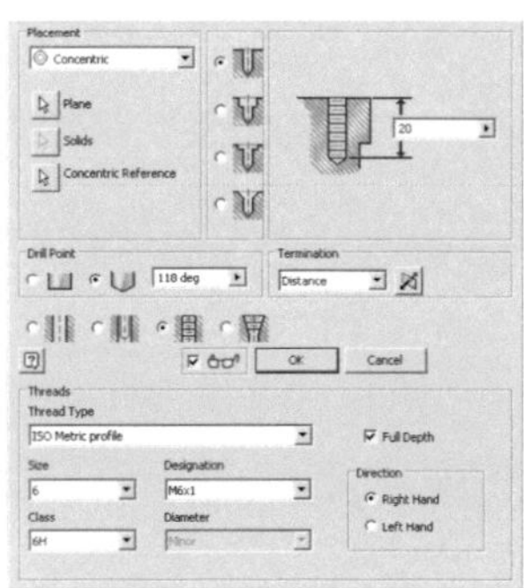

> ⬡ **Hole** (with thread)
> Placement: Concentric
> Termination: Distance 20
> Thread Type: ISO Metric profile
> Size: 6 - Right hand
> Designation: M6 x 1
> Class: 6H
> Depth: Full
> ☐ OK

Note: If you are using **inch** units, please transform the metric ones to inch system.

The other 9 holes you then create via ⬚ **Rectangular Pattern** (Direction 1: Distance 80 with Quantity 5 and Direction 2: Distance 130 with Quantity 2). Round off the marked inside edges of the two partial areas. Make sure to round all adjoining edges along the outline.

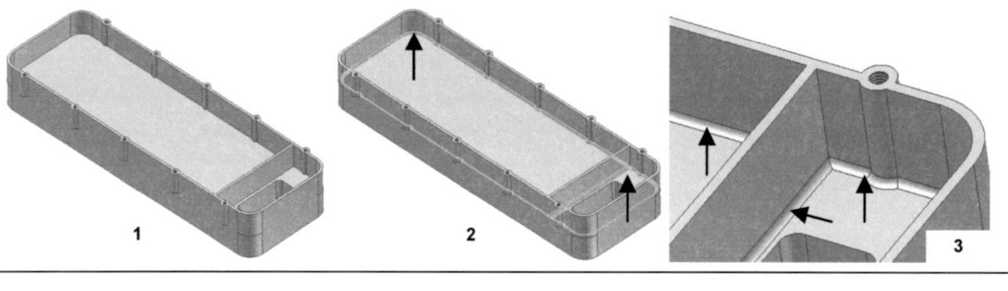

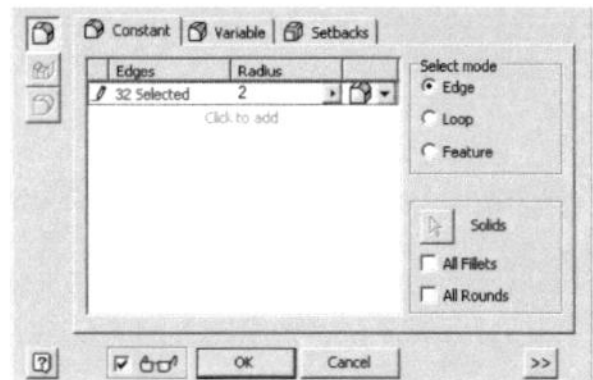

> ⬡ **Fillet**
> Edges: Marked edges (picture 2+3)
> Radius: 2
> ☐ OK

6.12.5 Receptions for the camshaft ring seal

To later seal the camshaft space towards the belt space we need a reception for a camshaft ring seal. Create a new ⬚ **Plane**, parallel to the marked wall (picture 1).

Then create a new ✎ **2D Sketch** on it and draw the shown outline.

Use the command ⬚ **Project Geometry**, to project the X-Y-Z-axes and upper wall edge.

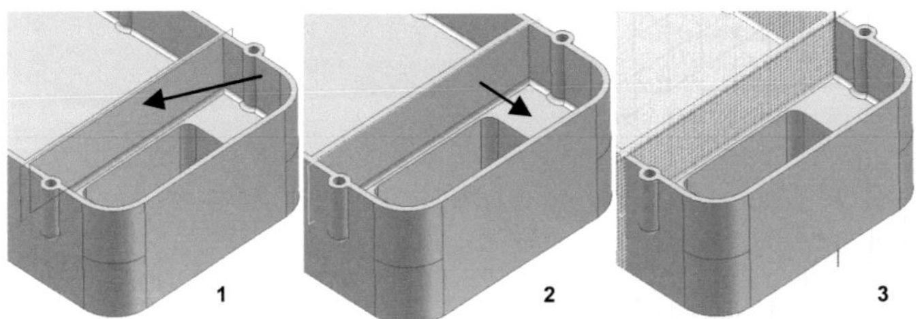

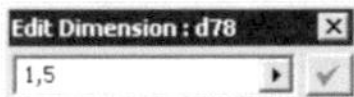

>  **Plane**
> Distance: 1.5 from face picture 1 (marked direction picture 2)
> ✔

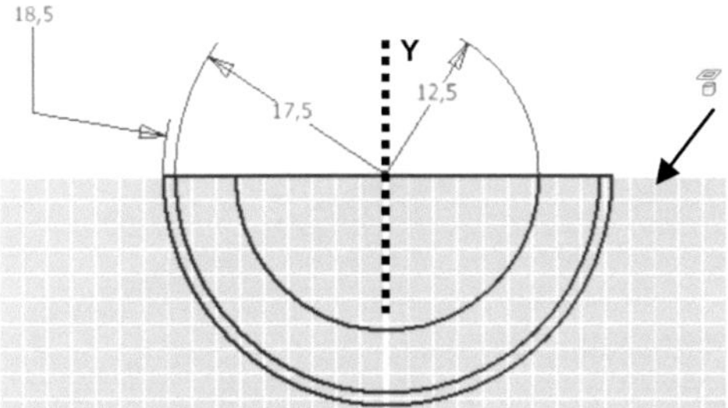

> 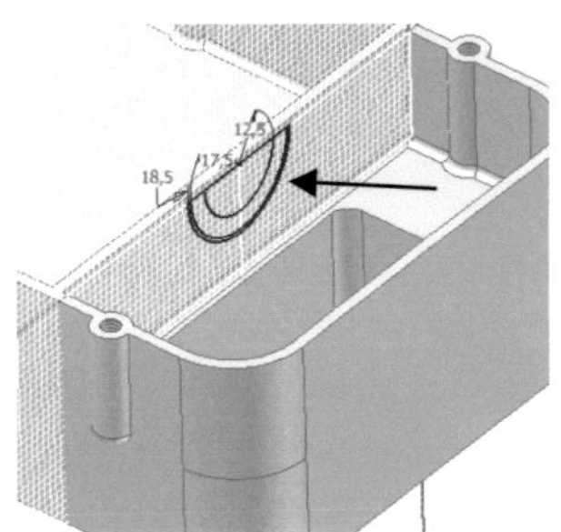 **2D Sketch** (on newly created plane)
> Slice sketch with **F7**
> **Project Geometry** (X-Y-Z-axes and marked edge)
> Draw shown geometry
> ✔ **Finish Sketch**

Execute following extrusions in 3 steps. After the first extrusion make the sketch visible again.

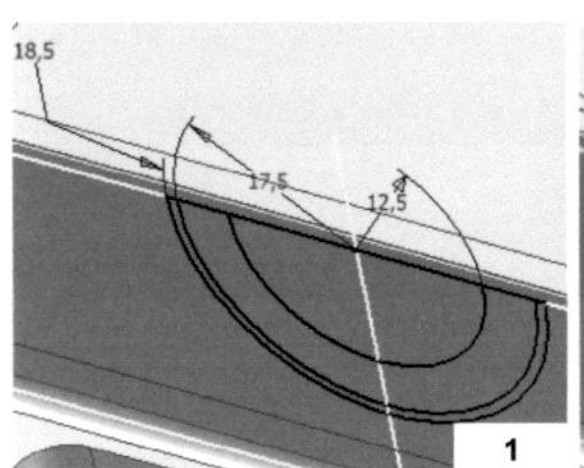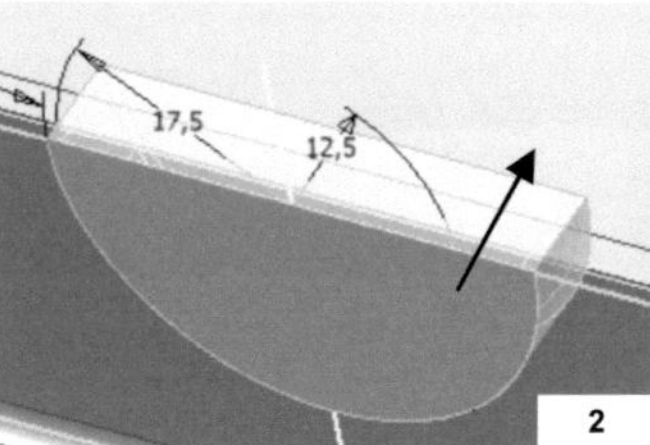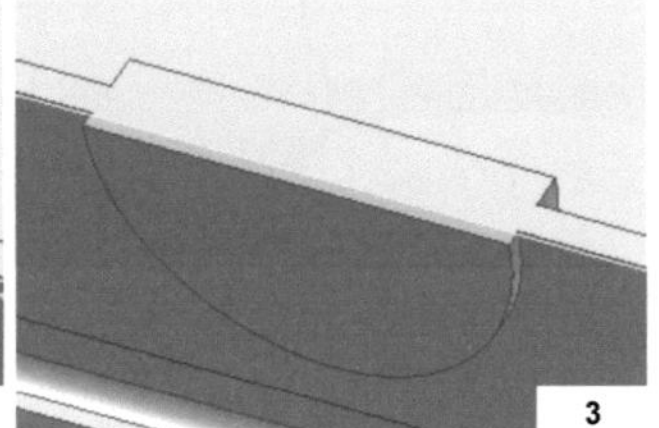

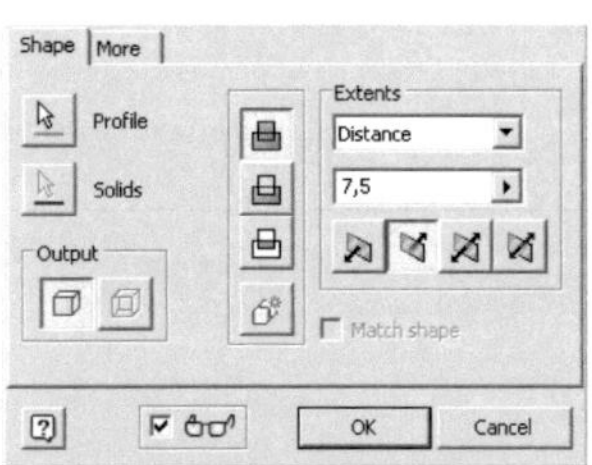

- ➢ 🗋 ***Extrude***
- ➢ Profile: Marked area (picture 2)
- ➢ Operation: Join
- ➢ Extents: Distance 7.5
- ➢ Direction: Shown (picture 2)
- ➢ [OK]

Tip: All 3 areas were extruded here!

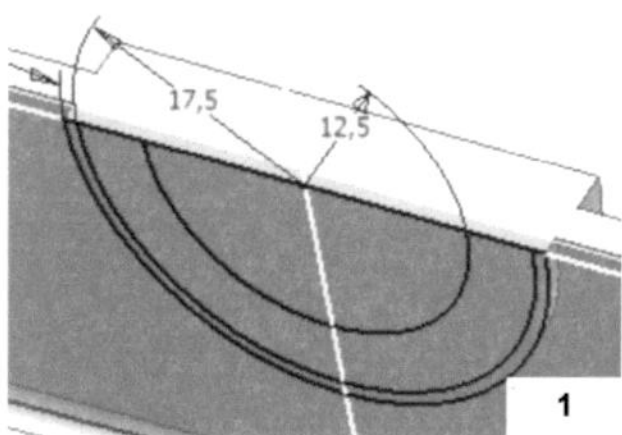
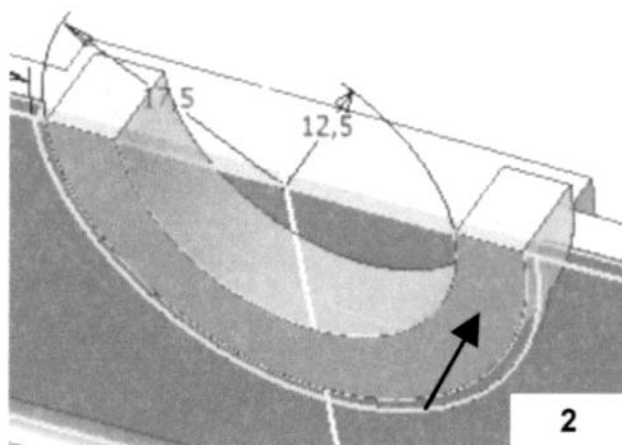
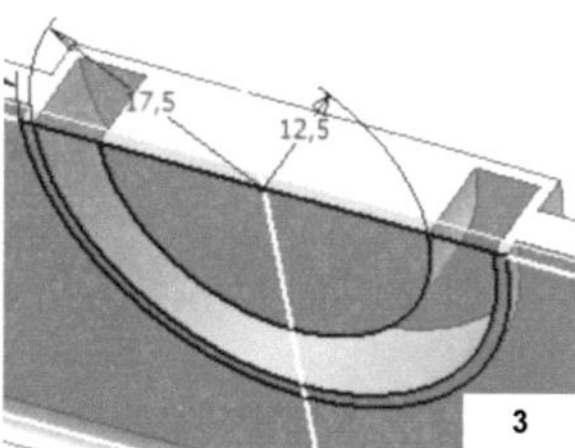

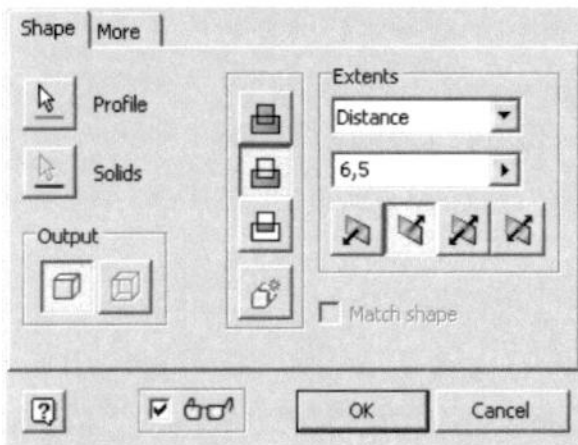

- ➢ 🗋 ***Extrude***
- ➢ Profile: Marked area (picture 2)
- ➢ Operation: Cut
- ➢ Extents: Distance 6.5
- ➢ Direction: Shown (picture 2)
- ➢ [OK]

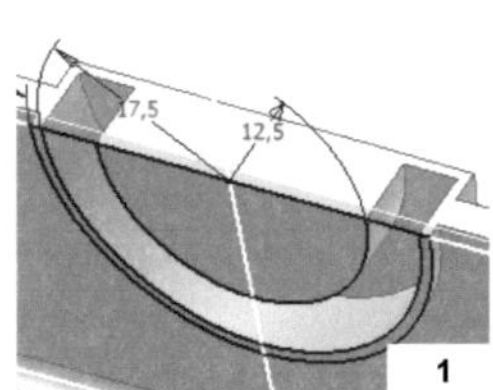
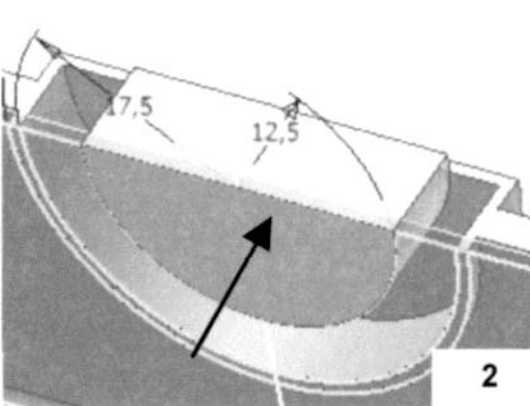
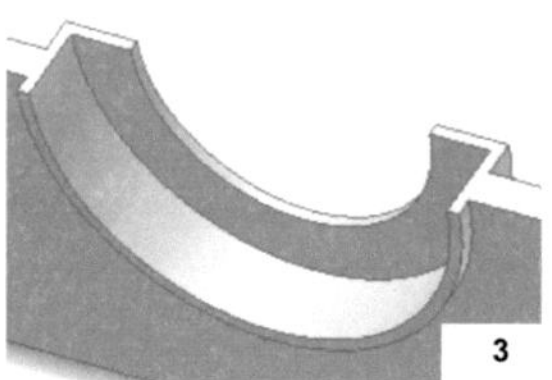

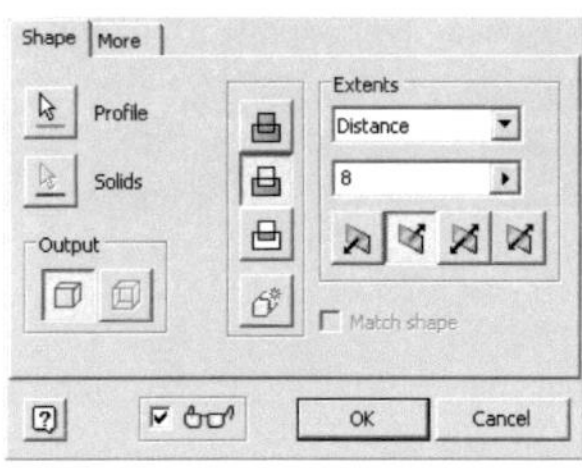

- ➢ 🗋 ***Extrude***
- ➢ Profile: Marked area (picture 2)
- ➢ Operation: Cut
- ➢ Extents: Distance 8
- ➢ Direction: Shown (picture 2)
- ➢ [OK]

6.12.6 Exhaust manifold of the cylinder space of the first cylinder

For the exhaust manifold for the fumes of the first cylinder we will create a mounting flange for an exhaust manifold.

We need a new ✎ **2D Sketch** on the marked wall (picture 1) and the 3 projected axes to be able to draw and dimension our outline. Then we extrude the outline.

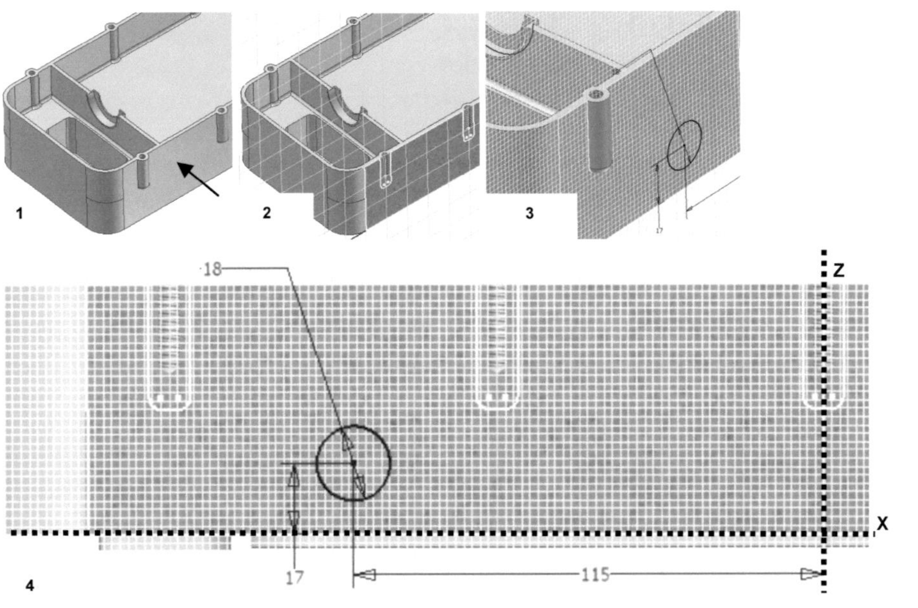

> ➤ ✎ **2D Sketch** (on marked area picture 1)
> ➤ ⁊ **Project Geometry** (X-Y-Z-axes)
> ➤ Draw shown circle
> ➤ ✔ **Finish Sketch**

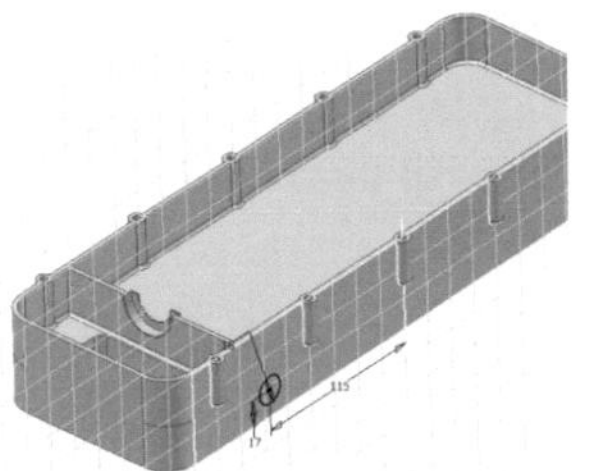

Tip: The dimension 115 refers to the distance center circle to Z-axis, the dimension 17 to the distance center circle to X- axis (picture 4).

Then extrude the volume shape from the sketch.

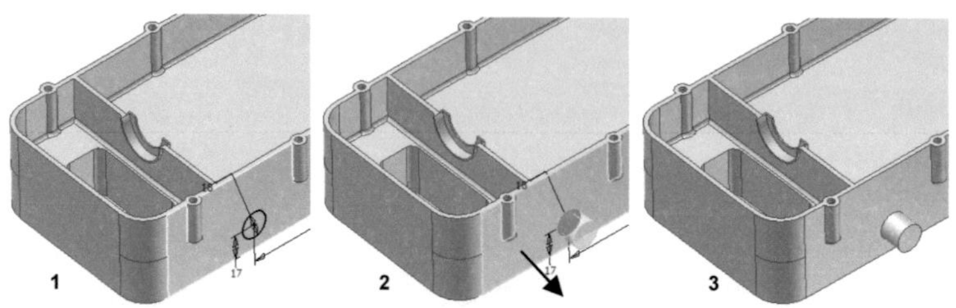

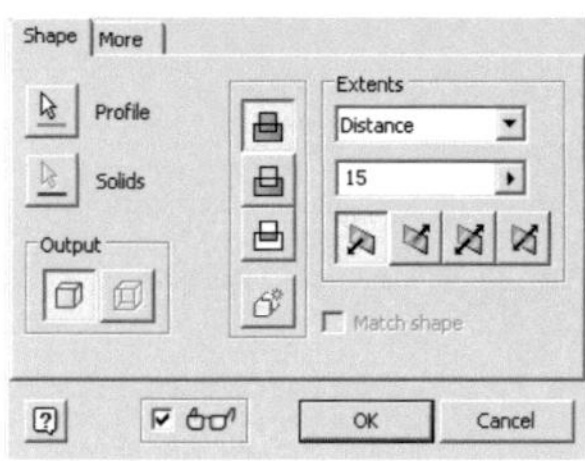

> 🗗 **Extrude**
> Profile: Circle
> Operation: Join
> Extents: Distance 15
> Direction: Shown (picture 2)
>   OK

For the connecting holes of the exhaust manifold between flange and cylinder area, which we will create with the command 🕉 **Sweep**, we need two sketches. The first 🗒 **2D Sketch** we can place on the face of the newly resulting cylinder.

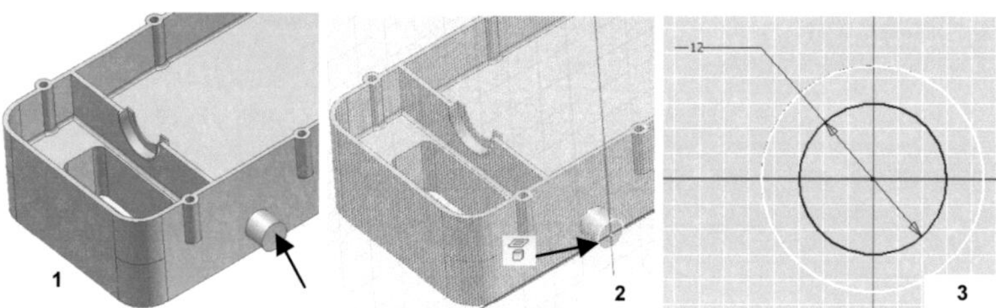

> 🗒 **2D Sketch** (on marked area picture 1)
> 🗗 **Project Geometry** (cylinder edge picture 2)
> Draw shown circle (D= 12, picture 3)
> ✔ **Finish Sketch**

Tip: The circle has to be located ◎ **concentric** to the projected cylinder edge.

For the second sketch we need a new 🗗 **Plane**, parallel to the Y-Z-plane. This we slice (**F7**) and draw the next shown outline, which later will be our extrude path.

Make sure that both end points of the line indeed close with the outline edges of the existing volume shape. Then we create a flow passage with the command 🕉 **Sweep**.

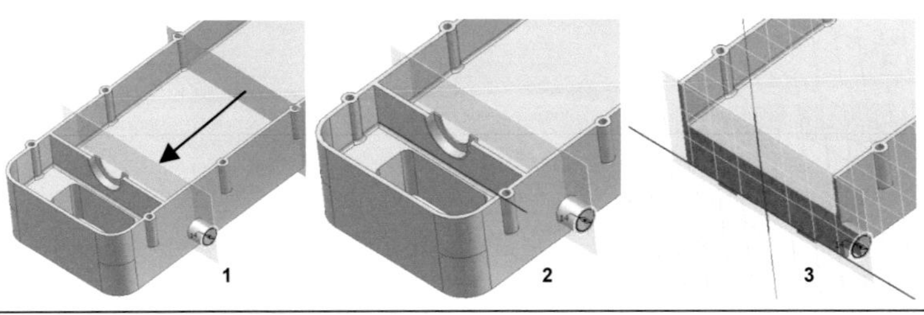

> ➢ ⬚ *Plane*
> ➢ Distance: 115 from Y-Z-plane (marked direction picture 1)
> ➢ ✔

Tip: The newly created plane now should go through the center of the existing cylinder.

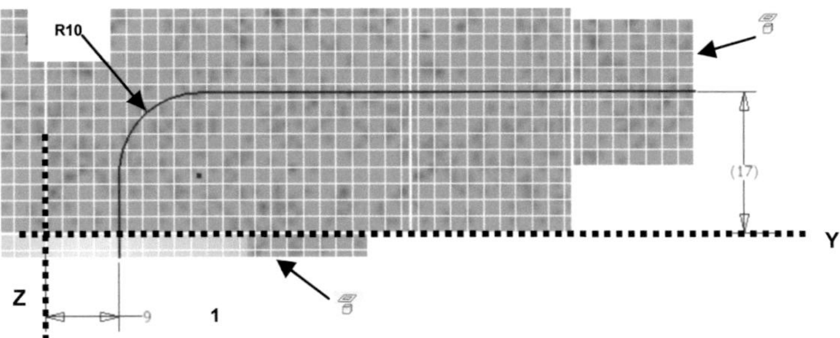

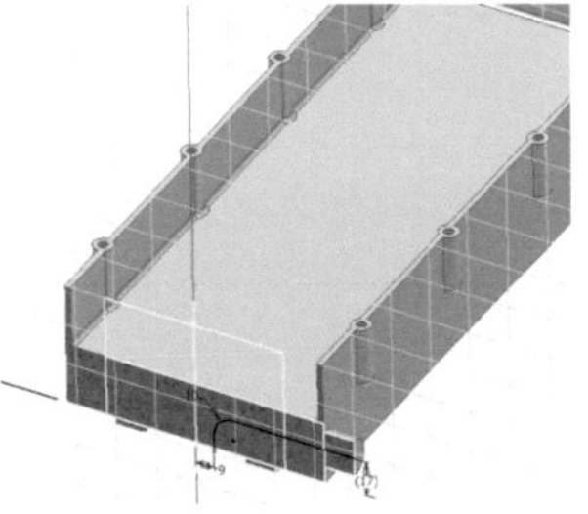

> ➢ ◫ *2D Sketch* (on newly created plane)
> ➢ ☰ *Project Geometry* (X-Y-Z-axes and both marked body edges)
> ➢ Draw shown outline
> ➢ ✔ *Finish Sketch*

Tip: Make sure that the two end points of the lines are precisely located on the projected outside edges.

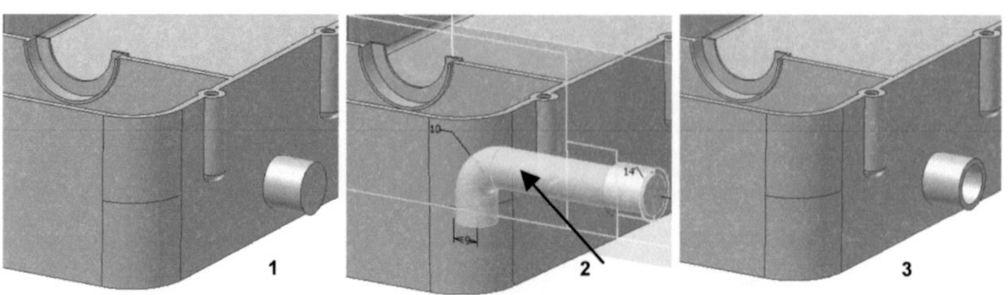

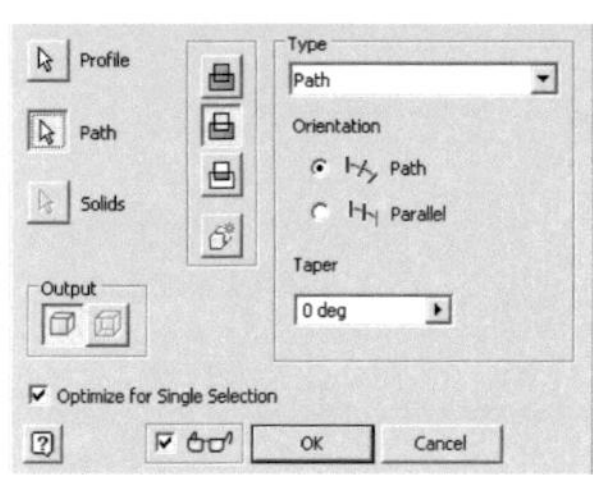

> 🜹 **Sweep**
> ➤ Profile: Circle from first sketch
> ➤ Path: Line from second sketch
> ➤ Type: Path
> ➤ Operation: Cut
> ➤ OK

Then we create a mounting flange for the exhaust manifold. Create a new ✎ **2D Sketch** on the marked circle area (picture 2) and project the <u>inner</u> existing circle ring (the outer one is not important here).

Draw the shown Ellipse, extrude and place following roundings.

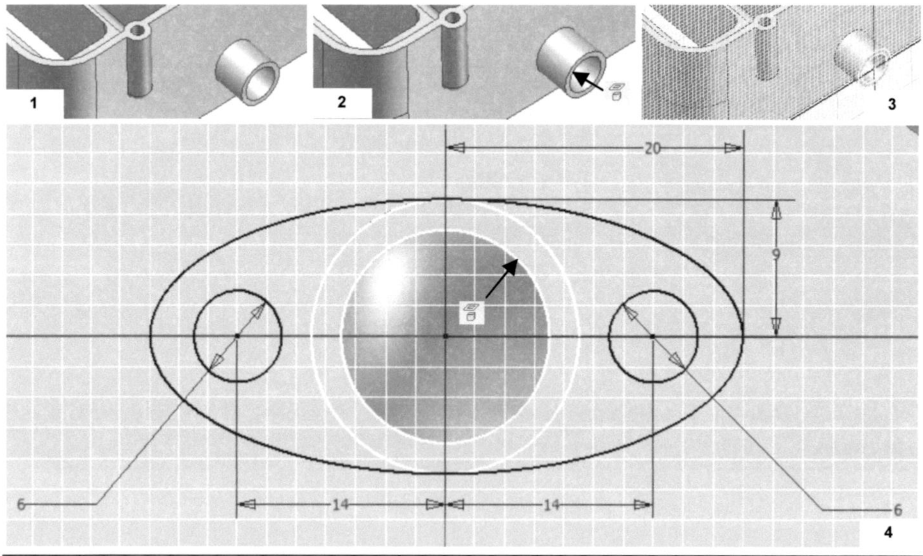

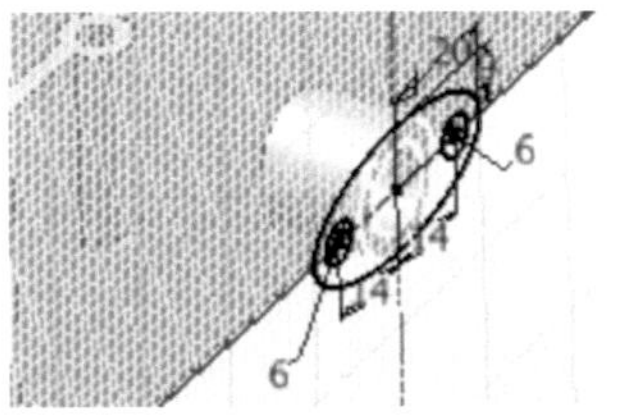

> ☑ **2D Sketch** (on marked circle area)
> ☞ **Project Geometry** (marked inner circle edge picture 4)
> Draw shown outline (1x ellipse and 2x circles)
> ✔ **Finish Sketch**

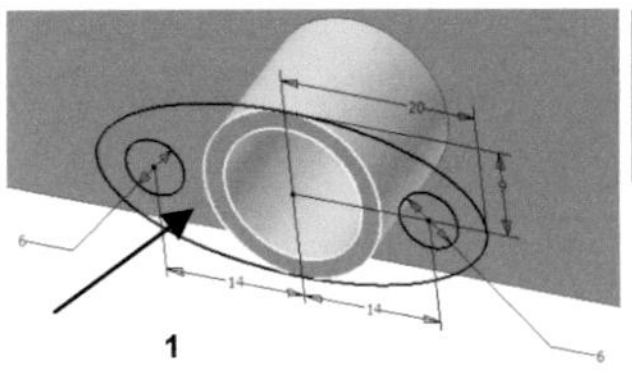

1

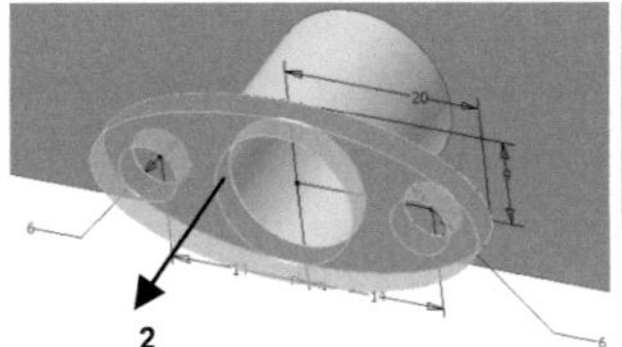

2

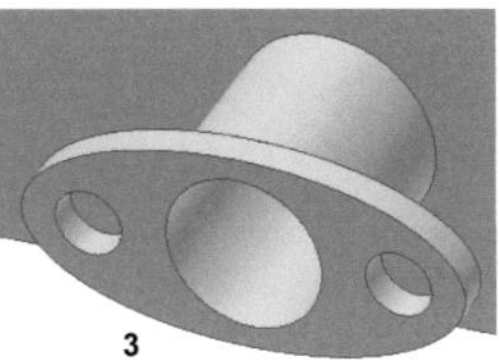

3

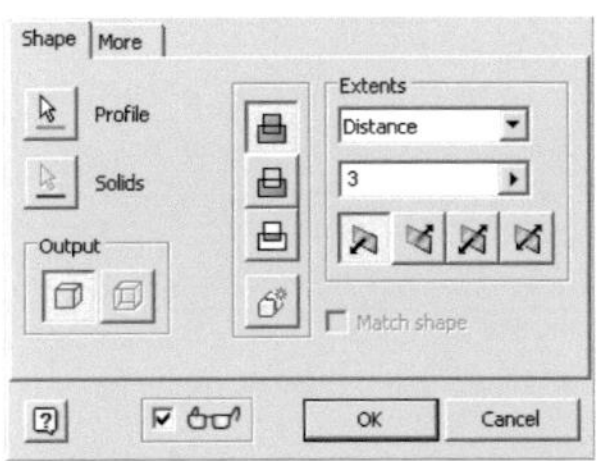

> ◫ **Extrude**
> Profile: Marked area (also the area between the inner circles, picture 1)
> Operation: Join
> Extents: Distance 3
> Direction: Shown (picture 2)
> OK

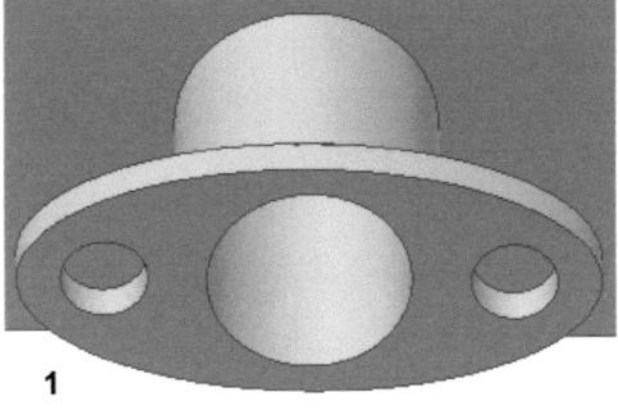

1

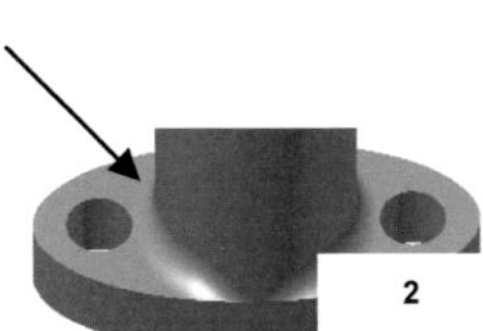

2

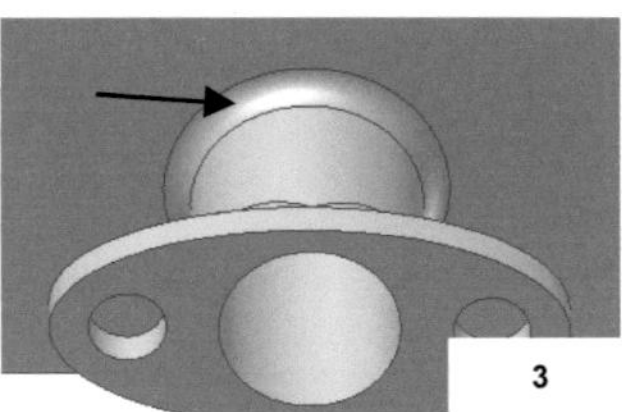

3

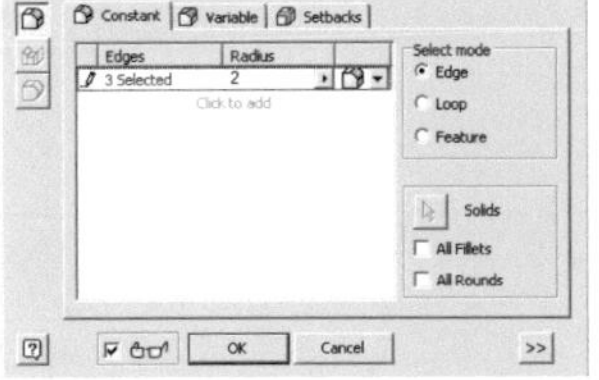

> ◔ **Fillet**
> Edges: 2 marked edges
> Radius: 2
> OK

6.12.7 Notch in the valve guide of the 1. cylinder

The valves are moved through the cams of the camshaft and this way open or close the intake and exhaust manifolds of the cylinder spaces.

Therefore we need a connective hole between camshaft and cylinder space.

Create a new ✎ **2D Sketch** on the marked seal area (picture 1), project the marked existing edges as well as the X-Y-Z-axes and then draw the shown two circles (D1= 3 and D2= 16) in the center of the projected manifold (picture 3).

Then create a material cut for the valve guide with the small circle.

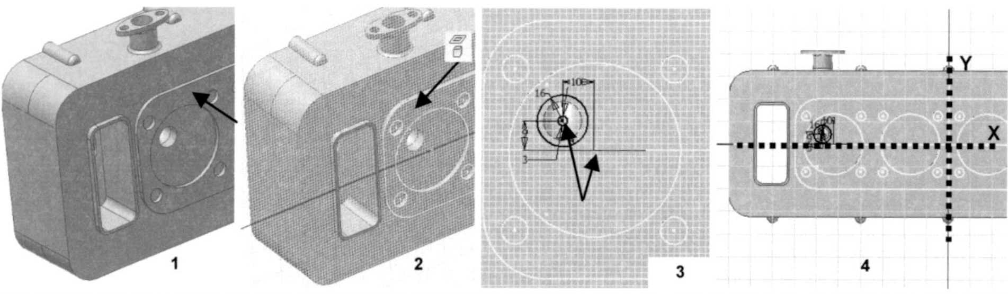

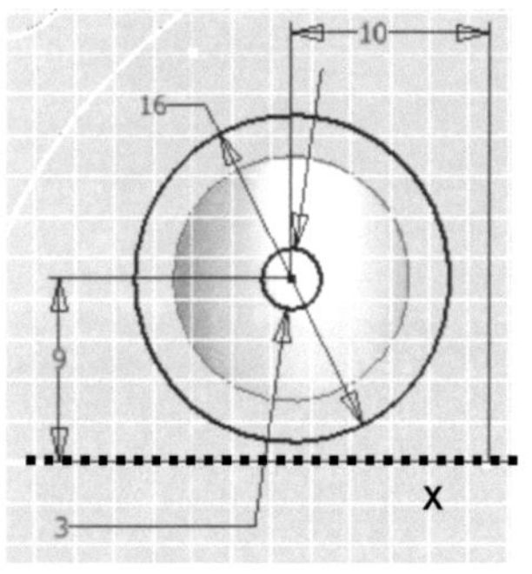

> ✎ **2D Sketch** (on marked area picture 1)
> ▸ **Project Geometry** (X-Y-Z-axes and marked area with all edges in picture 2)
> ▸ Draw shown circles (D1= 3, D2= 16)
> ✔ **Finish Sketch**

Tip: The dimensions 9 and 10 refer to the distances between the two in picture 3 marked centers of the circles.

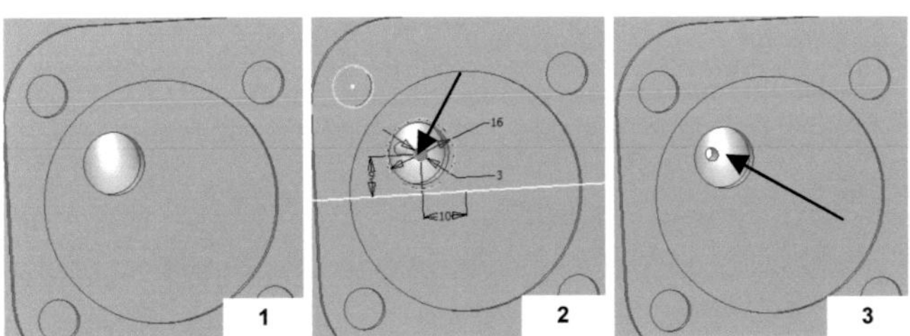

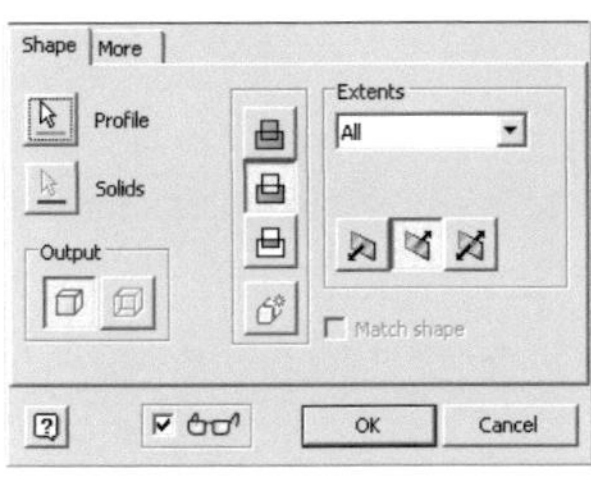

> ◻ **Extrude**
> Profile: Circle (D=3)
> Operation: Cut
> Extents: All
> Direction: Shown (picture 2)
> <u>OK</u>

6.12.8 Through-holes for cylinder head screws

For the through-holes we use the previously used sketch (***RMC*** > ***Share Sketch***).

Create the 4 marked holes (picture 1) by using the command ◻ ***Extrude***.

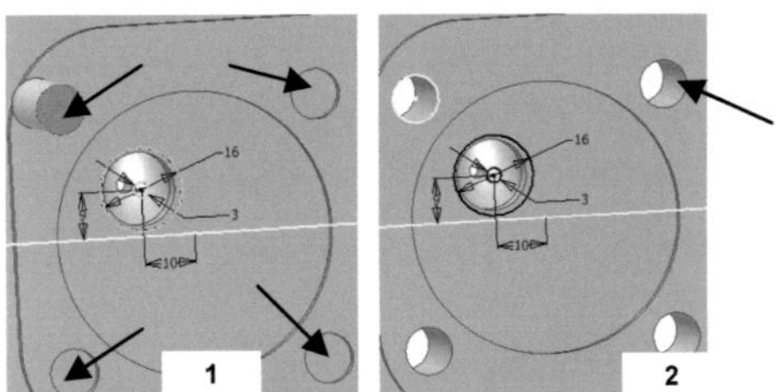

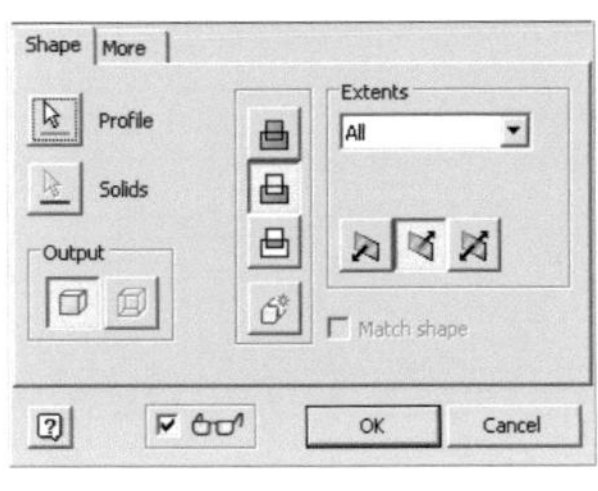

> ◻ **Extrude**
> Profile: 4 holes (picture 1)
> Operation: Cut
> Extents: All
> Direction: Shown (picture 2)
> <u>OK</u>

6.12.9 Creation of the seal area of the valve seat

The seal area of the valve seat requires another material cut. Use the still activated sketch and extrude the circle (D= 16).

Then create a ⬡ **Chamfer**, which later will form the seal area between combustion chamber space and valve head.

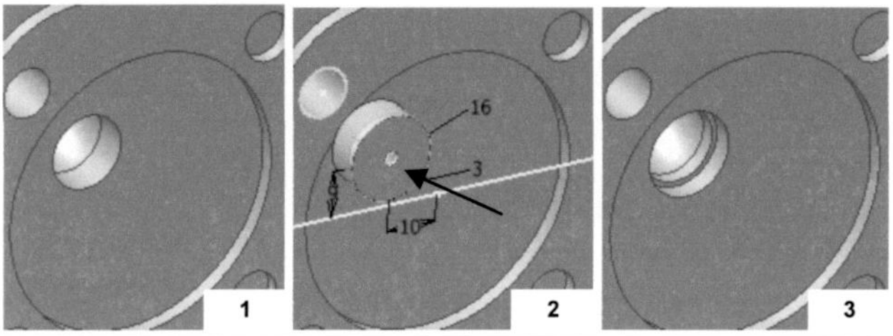

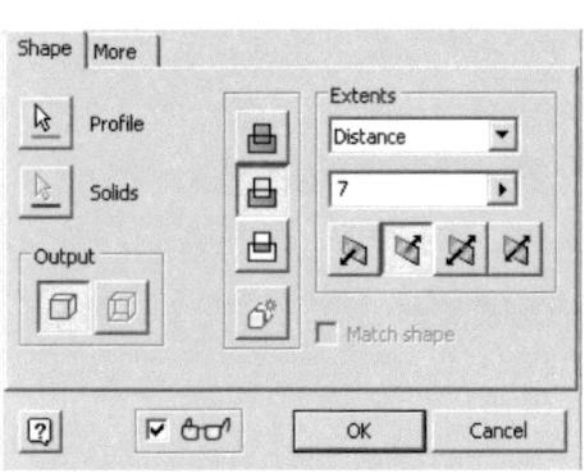

> ⬛ **Extrude**
> ➤ Profile: Circle (D= 16)
> ➤ Operation: Cut
> ➤ Extents: Distance 7
> ➤ Direction: Shown (picture 2)
> ➤ OK

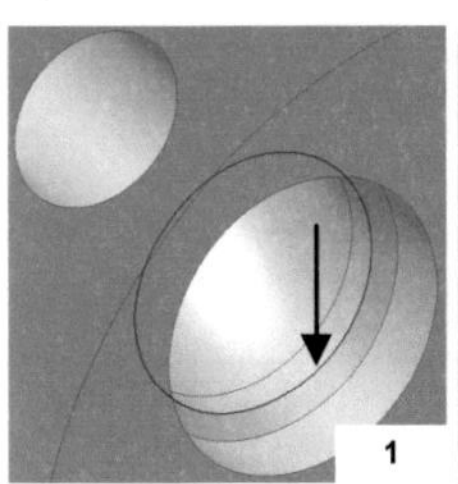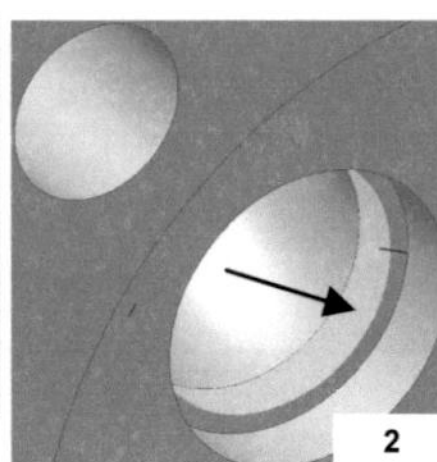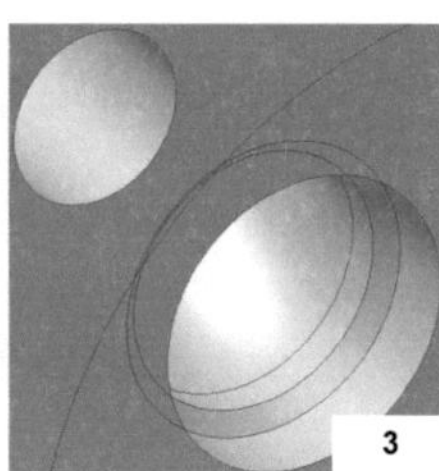

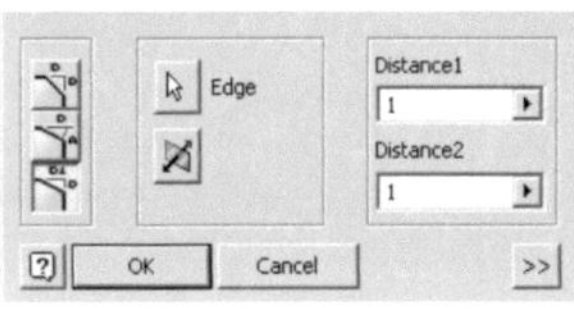

> ⬡ **Chamfer**
> ➤ Edge: Inner edge of the hole
> ➤ Distance_1: 1
> ➤ Distance_2: 1
> ➤ OK

6.12.10 Intake manifold & holes from the first cylinder

With the command ⊕ *Circular Pattern* we copy the marked elements (picture 1).

Use the circle edge of the cylinder space as rotation axis (picture 3).

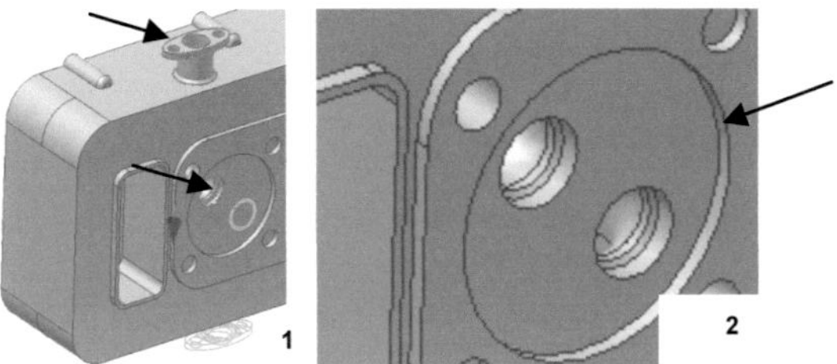

> ⊕ *Circular Pattern*
> Elements: All valve and manifold elements (beside the Through-holes from chapter 6.12.8)
> Axis: Marked circle edge (picture 2)
> Quantity: 2
> Rotation: 360°
>   OK

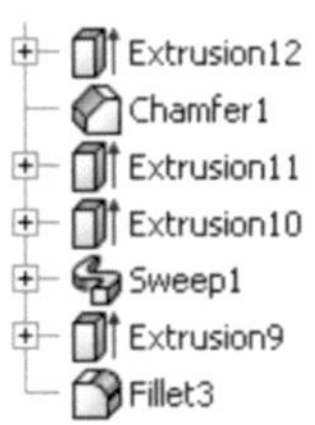

Then round the marked edge (picture 2+3) in the first cylinder.

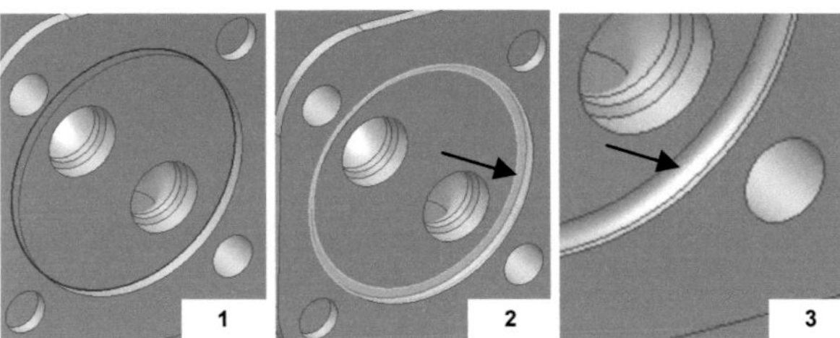

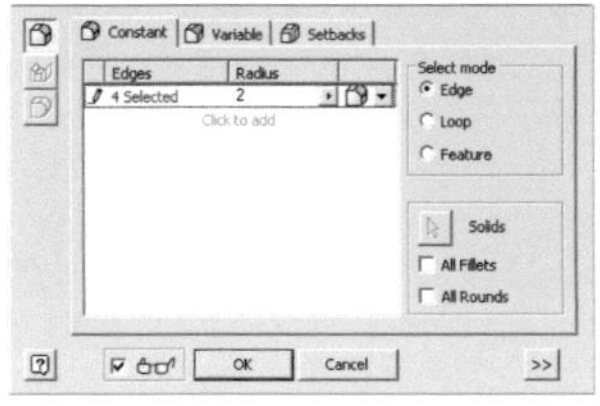

> ◯ *Fillet*
> Edges: Marked edge (picture 2+3)
> Radius: 2
>   OK

6.12.11 Camshaft bearing from the first cylinder

For the camshaft bearing above the first cylinder create a ✎ **2D Sketch** on the marked area and then draw the shown outline.

Project the X-Y-Z-axes and the 4 marked bore edges.

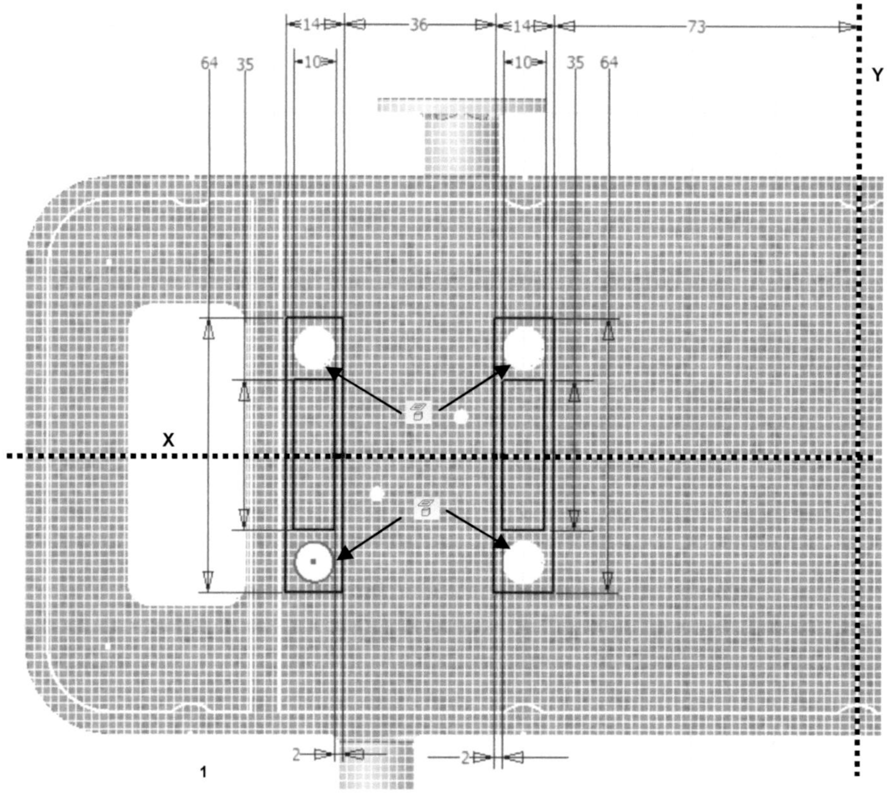

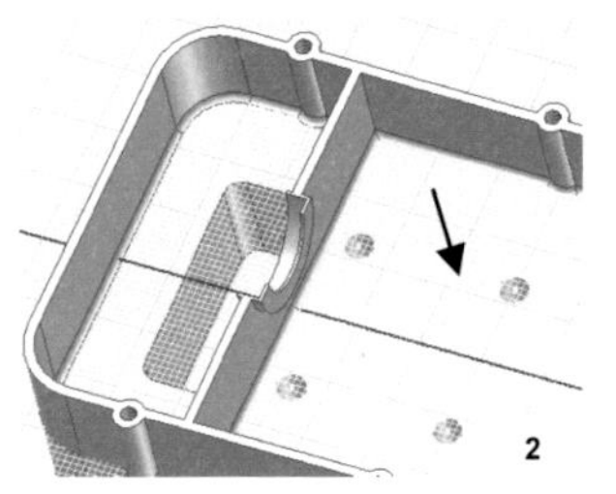

> ✎ **2D Sketch** (on marked area picture 2)
> ⚙ **Project Geometry** (X-Y-Z-axes and marked circle-edges in picture 1)
> ➢ Draw shown four rectangles
> ✔ **Finish Sketch**

Now create a volume shape from the two inner rectangles (markings in picture 1).

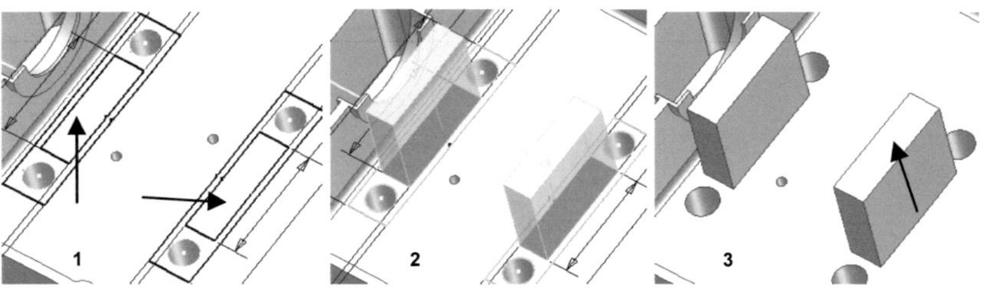

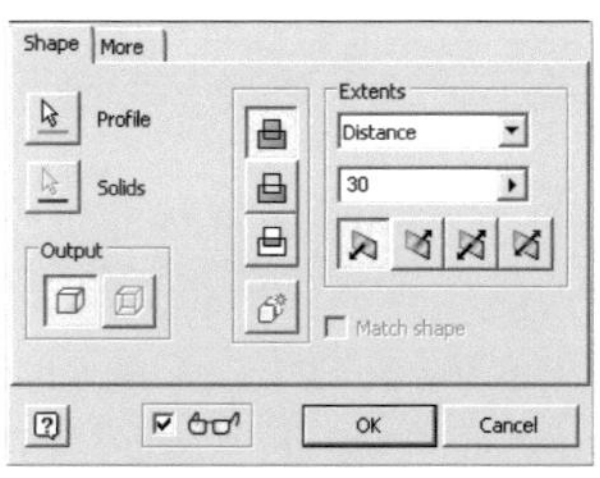

> 🗍 ***Extrude***
> ➢ Profile: Marked areas (picture 1)
> ➢ Operation: Join
> ➢ Extents: Distance 30
> ➢ Direction: Shown (picture 3)
> ➢   OK

Now make the sketch visible again and extrude the outer rectangles (without the 4 circles). Round off the outside edges of the created volume shapes (all in all 40 edges).

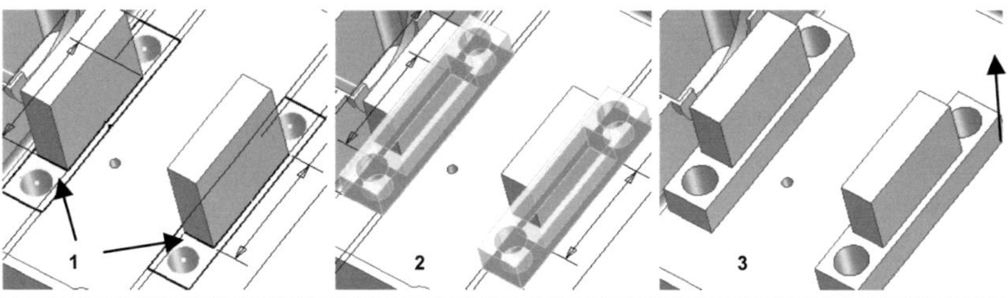

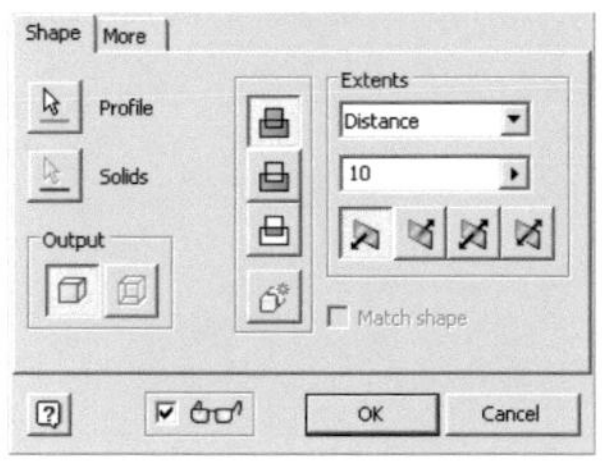

> 🗍 ***Extrude***
> ➢ Profile: Marked areas (picture 1)
> ➢ Operation: Join
> ➢ Extents: Distance 10
> ➢ Direction: Shown (picture 2)
> ➢   OK

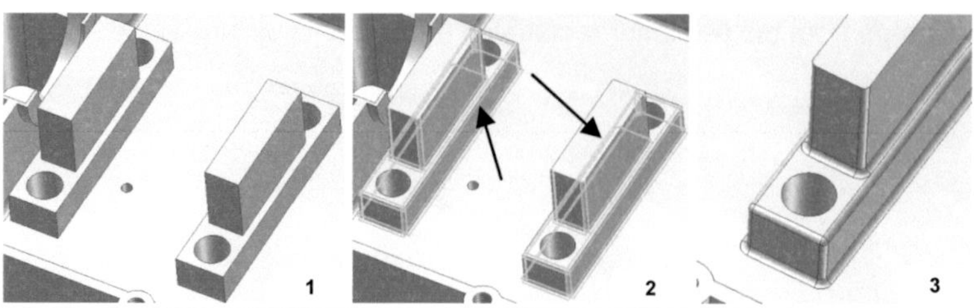

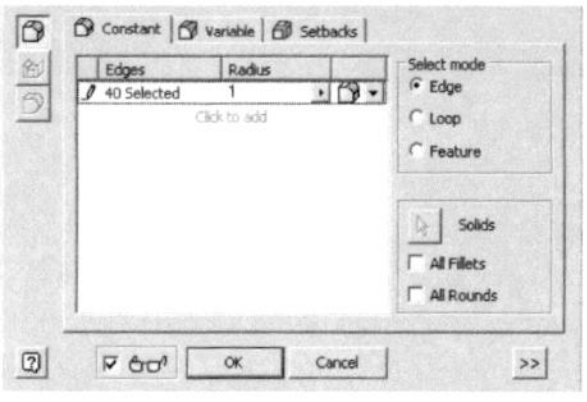

> ⬡ **Fillet**
> Edges: All 40 edges of the four new bodys (beside the four hole- edges)
> Radius: 1
> [OK]

On the marked area in picture 1 create a new ✎ **2D Sketch**, project the X-Y-Z-axes as well as the marked edge shown in picture 3 and draw the following half circle (R= 10). Then extrude the material cut.

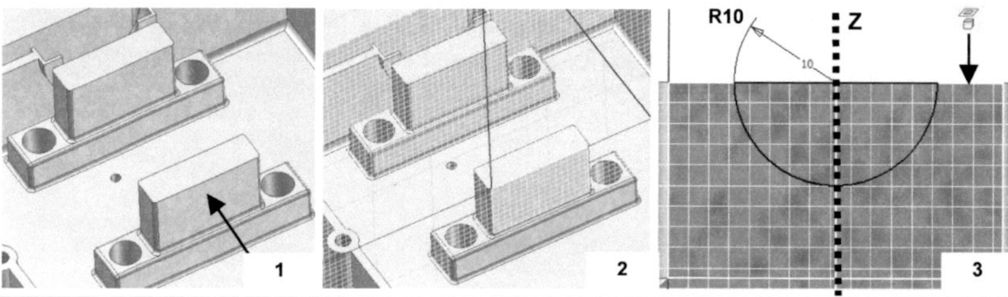

> ✎ **2D Sketch** (marked area picture 1)
> ✇ **Project Geometry** (X-Y-Z-axes and marked edge picture 3)
> Draw shown outline
> ✔ **Finish Sketch**

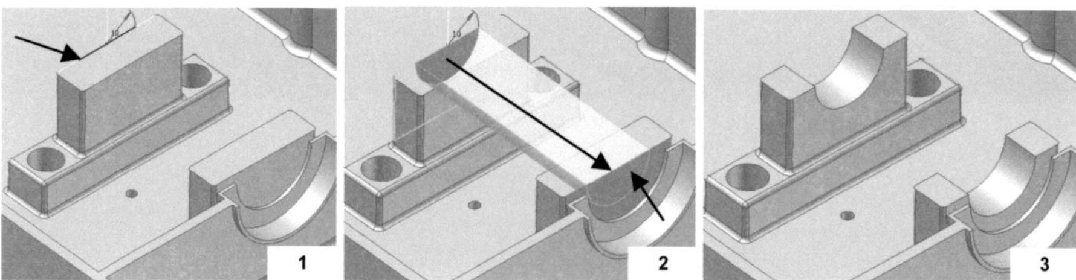

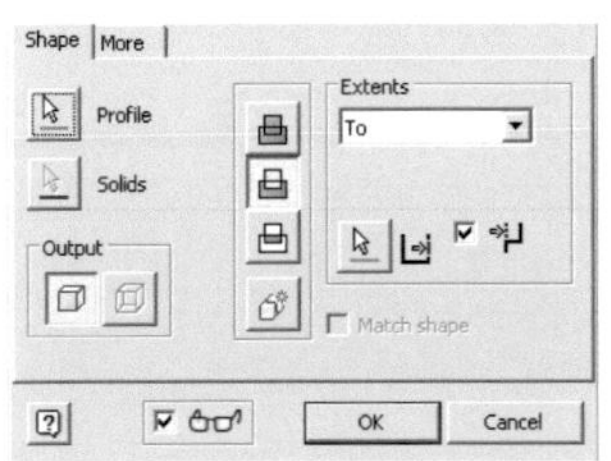

> 🗔 **Extrude**
> ➢ Profile: Area of last sketch
> ➢ Operation: Cut
> ➢ Extents: To
> ➢ To: Marked face in picture 2
> ➢ [OK]

Tip: The half circle of the previously drawn sketch is located on the backside of the outline in picture 1 (see arrow).

6.12.12 Spark plug hole from the first cylinder

Our engine has a camshaft which is positioned in the center of the cylinder head.

Therefore our spark plugs have to get screwed into the cylinder head in a 20° angle.

For this we need a 🗐 **Plane** twisted by this angle on which then a new 🖉 **2D Sketch** is created. Then draw the 3 shown circles and place the ✛ **Point** in the middle.

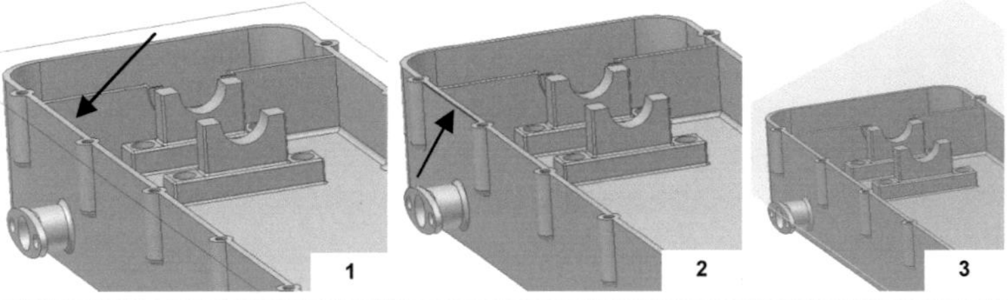

> ➢ 🗐 **Plane**
> ➢ The upper seal area of the body (picture 1) and then click the rotation edge (picture 2)
> ➢ Adjust the inclination to 20°
> ➢ ✔

Tip: Should the plane rotate into the wrong area try entering a negative algebraic sign.

Use the command 🖮 **View Face** and click the new plane, to adjust the drawing view to the tilted plane.

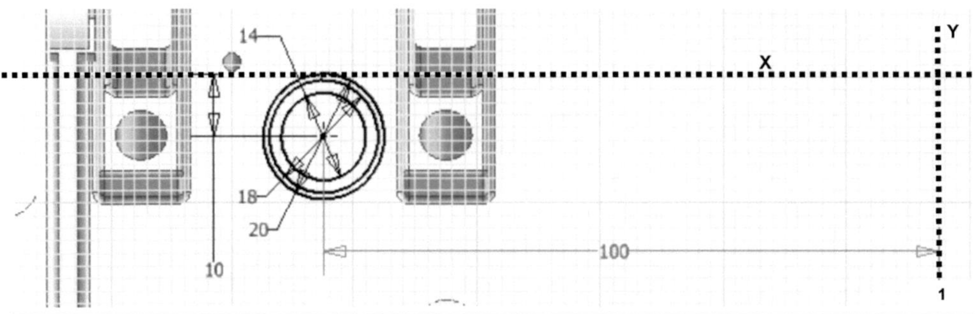

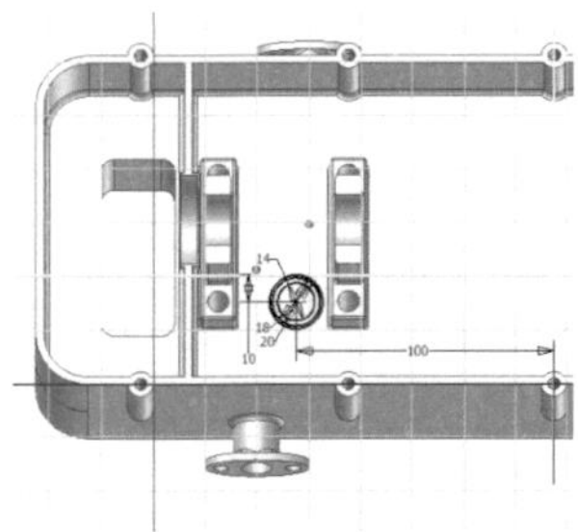

> 🖉 **2D Sketch** (on newly created plane)
> 🖩 **Project Geometry** (X-Y-Z-axes)
> Draw the shown three circles (D1= 14, D2= 18, D3= 20) and ✛ **Point** in the middle
> ✔ **Finish Sketch**

First create a 🔩 **Threaded hole**. Make the sketch visible again and execute following 3 extrusions. Make sure to select the correct areas!

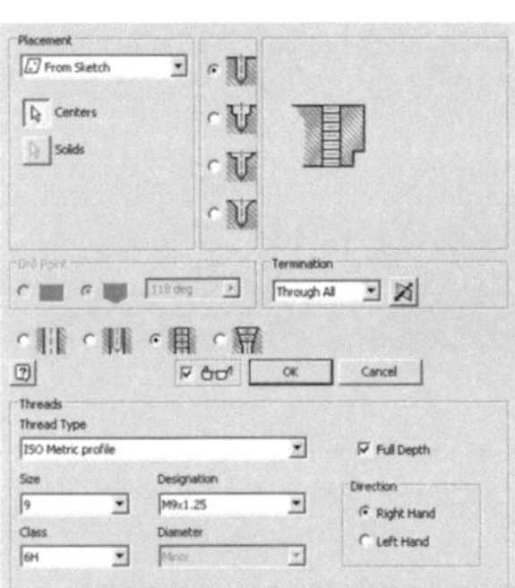

> 🔩 **Hole** (with thread)
> Placement: From Sketch
> Termination: Through All
> Thread Type: ISO Metric profile
> Size: 9 - Right hand
> Designation: M9 x 1.25
> Class: 6H
> Depth: Full
> ☐ OK

Note: If you are using **inch** units, please transform the metric ones to inch system.

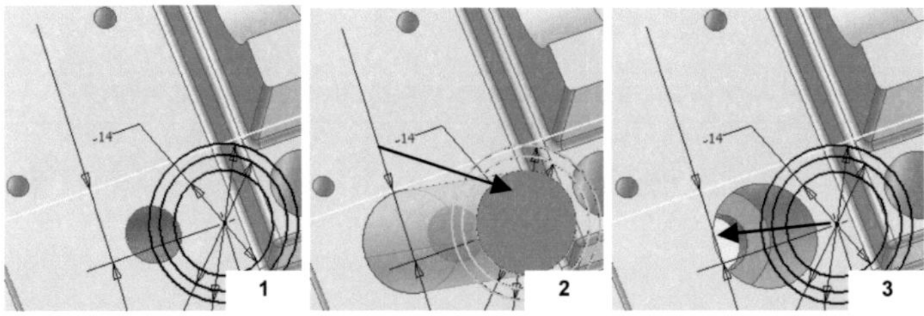

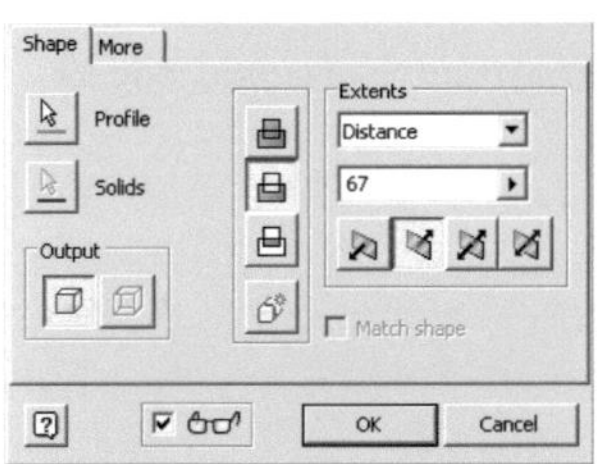

➤  ***Extrude***
➤ Profile: Circle (D= 14) (picture 2)
➤ Operation: Cut
➤ Extents: Distance 70
➤ Direction: Shown (picture 3)
➤ OK

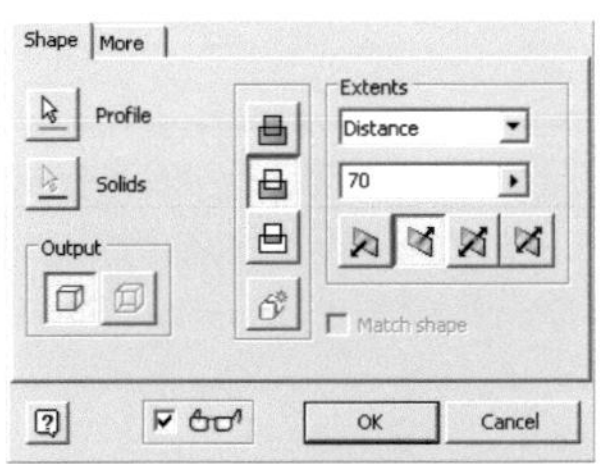

➤ ***Extrude***
➤ Profile: Marked circle area between D= 14 and D= 18 (picture 2)
➤ Operation: Cut
➤ Extents: Distance 67
➤ Direction: Shown (picture 3)
➤ OK

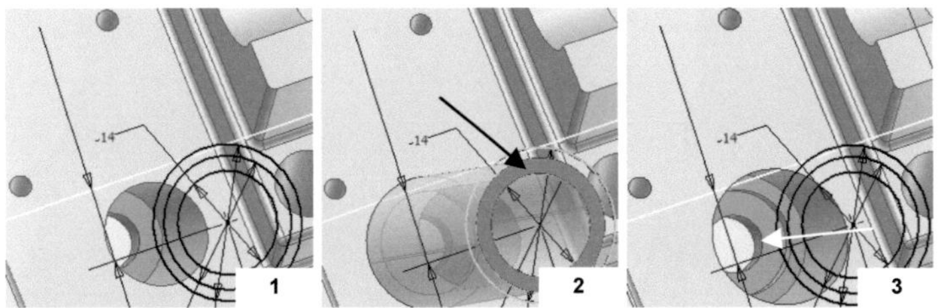

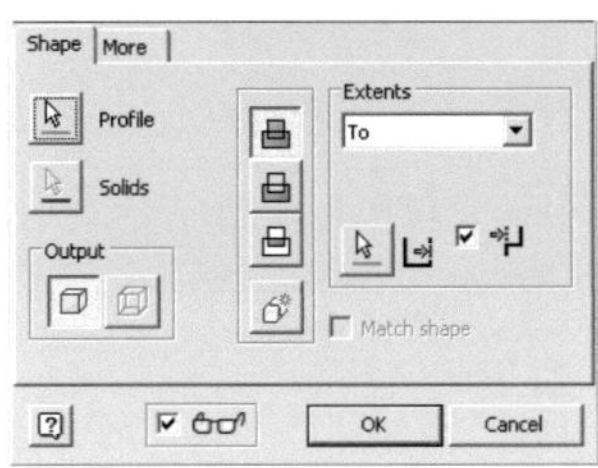

➤ ***Extrude***
➤ Profile: Marked circle area between D= 18 and D= 20 (picture 2)
➤ Operation: Join
➤ Extents: To
➤ To: Marked face (picture 3)
➤ OK

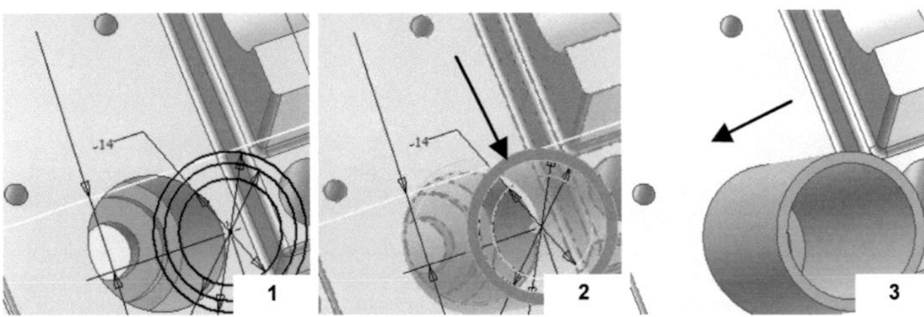

6.12.13 Threaded holes for camshaft screws

We yet need ⊚ **Threaded holes** for the camshaft screws. For this you have to create a new ✑ **2D Sketch** on the marked area and place 4 ╬ **Points** there. Project the marked edges (two are marked in picture 1).

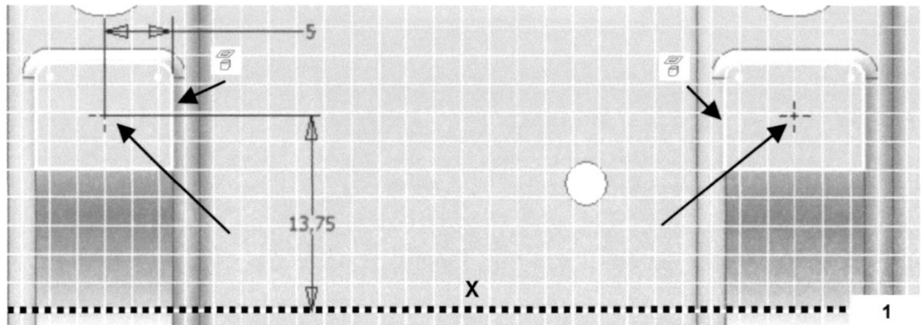

> ✑ **2D Sketch** (on marked plane)
> ⊟ **Project Geometry** (X-Y-Z-axes and marked edges)
> Place 4 ╬ **Points** as shown
> ✔ **Finish Sketch**

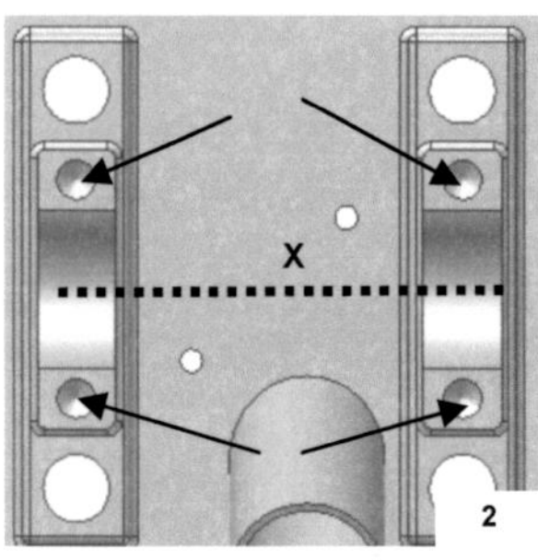

Alternative you can genetate the 4 points by fix the first one and then using ⊟ **Rectangular Pattern** to create the other 3. Therefor use the follow settings:

> 2x 50 with direction X-axis and
> 2x 27.5 with direction Y-axis

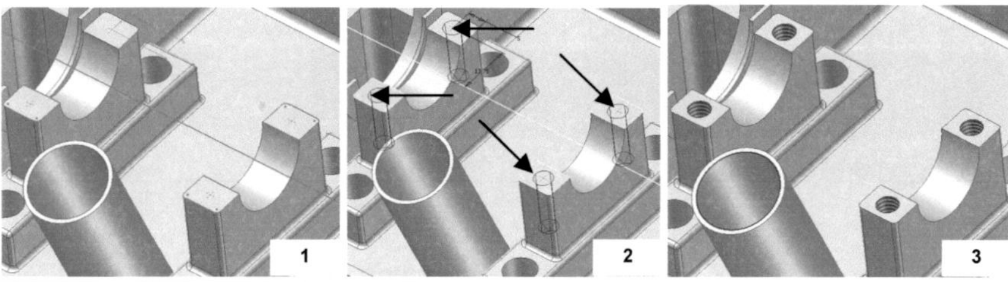

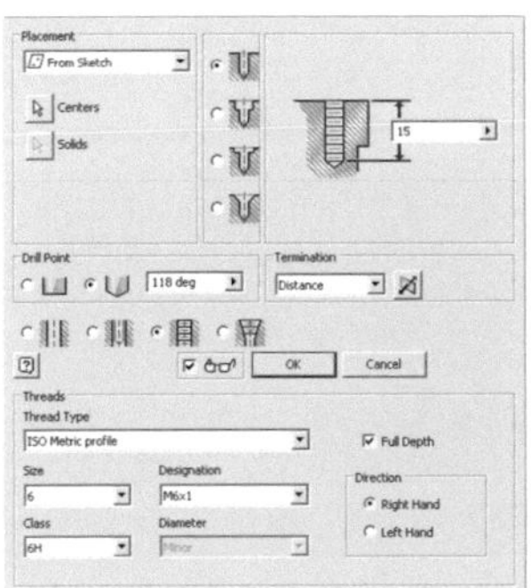

> ⬡ **Hole** (with thread)
> Placement: From Sketch
> Termination: Distance 15
> Thread Type: ISO Metric profile
> Size: 6 - Right hand
> Designation: M6 x 1
> Class: 6H
> Depth: Full
> [OK]

Note: If you are using *inch* units, please transform the metric ones to inch system.

6.12.14 Create elements for cylinder 2-4

It is time to adopt all the elements created for the first cylinder also for the other 3 cylinders.

Use the command ⬚ **Rectangular Pattern**, to copy all elements modeled on the first cylinder.

Make sure to mark indeed all necessary elements within the model tree.

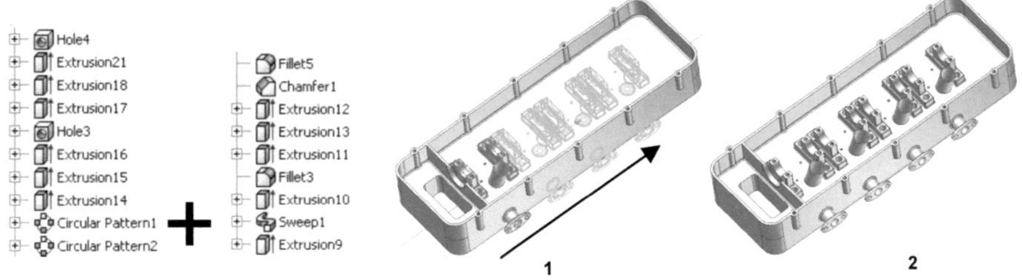

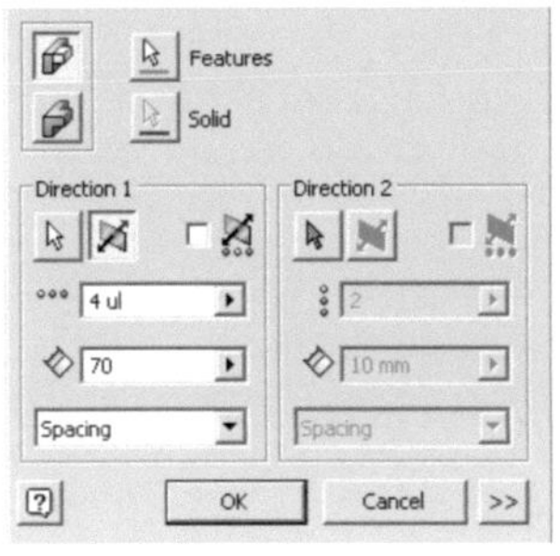

> ⬚ **Rectangular Pattern**
> Elements: Shown elements (picture 1)
> Direction: X- axis
> Quantity: 4
> Distance: 70
> [OK]

6.12.15 Beveling the valve exhaust edges

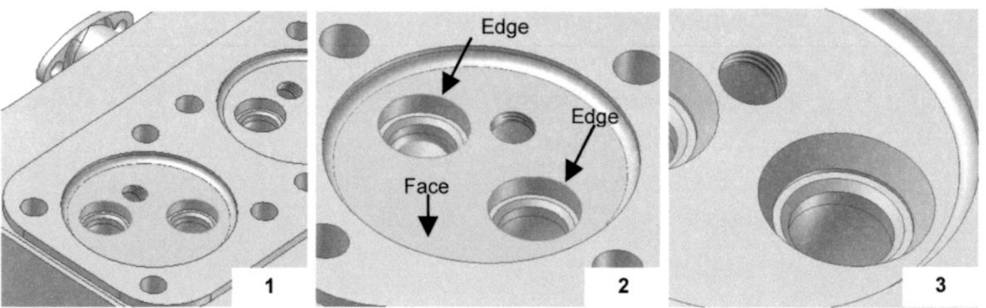

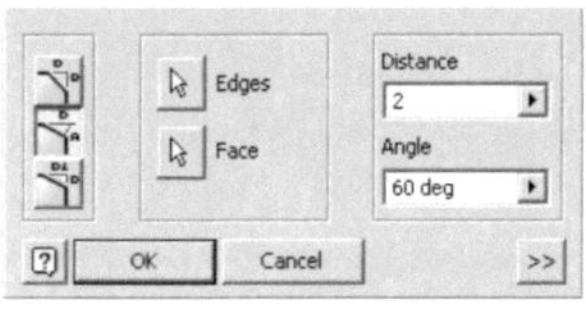

> ◇ **Chamfer**
> ➢ Edges: Marked 2 edges (picture 2)
> ➢ Face: Shown floorspace (picture 2)
> ➢ Distance: 2
> ➢ Angle: 60°
> ➢ OK

Repeat the beveling on the remaining 6 valve seats.

Tip: The beveling of the 8 valve seats was purposefully pushed back and not used together with the command ⊞ **Rectangular Pattern**. Unfortunately there are problems creating the bevels if this order is not kept.

cylinder-head.ipt now is complete. 💾 **Save** and ✖ **Close** the file.

6.13 Part CAMSHAFT
6.13.1 Basics of the part

The job of the camshaft lobes is to change the rotational movement of the camshaft into a linear movement of the valves. The shaft is powered by a belt connection to the crankshaft.

Execution and order of the cams determine a defined guidance of the intake and exhaust manifolds.

6.13.2 Base body by extrusion

First we will create a simple lobe. Create a ⬚ **New Part** and 💾 **Save** it as **camshaft.ipt** in your project folder. Project the X-Y-Z-axes and draw a circle (D= 20) in the coordinate's origin. Use the command ⬚ **Extrude** to then create the volume shape.

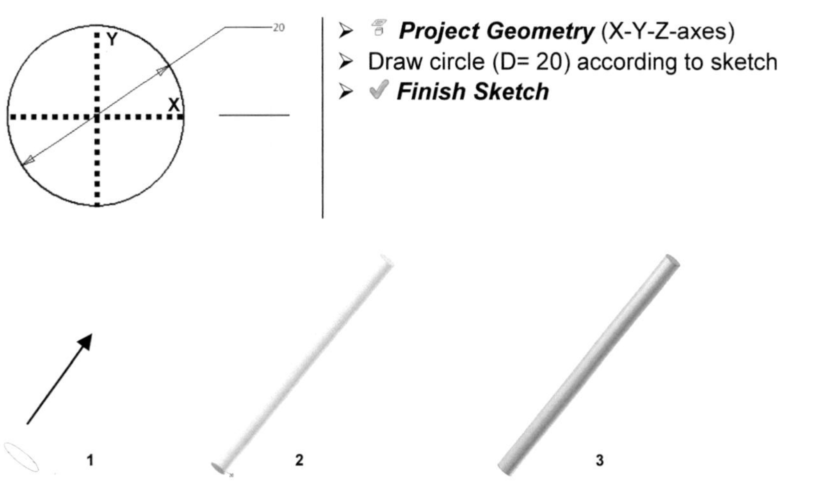

> ➤ 🖫 ***Project Geometry*** (X-Y-Z-axes)
> ➤ Draw circle (D= 20) according to sketch
> ➤ ✔ ***Finish Sketch***

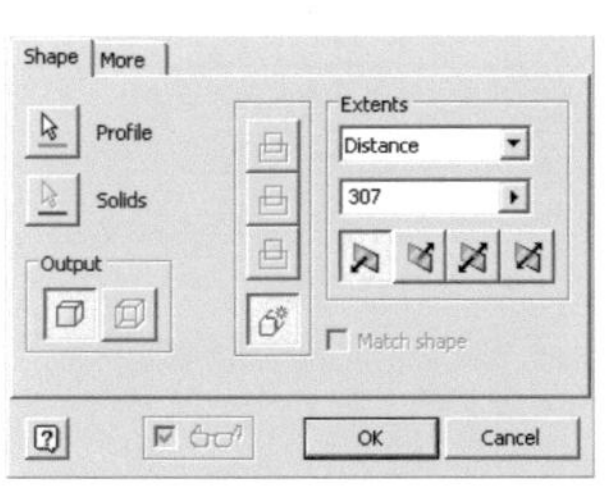

> ➤ ▯ ***Extrude***
> ➤ Profile: Circle
> ➤ Operation: Join
> ➤ Extents: Distance 307
> ➤ Direction: Shown (picture 1)
> ➤ ☐ OK

6.13.3 Guides of the camshaft

Our camshaft is guided in the cylinder head by 8 bearings. They should enable the rotation movement and prevent a distance in axial direction. For this we need 16 guides on the camshaft (2 per bearing). Create a new 🖉 ***2D Sketch*** on the Y-Z-plane and then draw the 4 shown rectangles (2 x 2). Project all necessary axes and edges. Then use the command 🖴 ***Revolve***, to generate the 4 volume shapes.

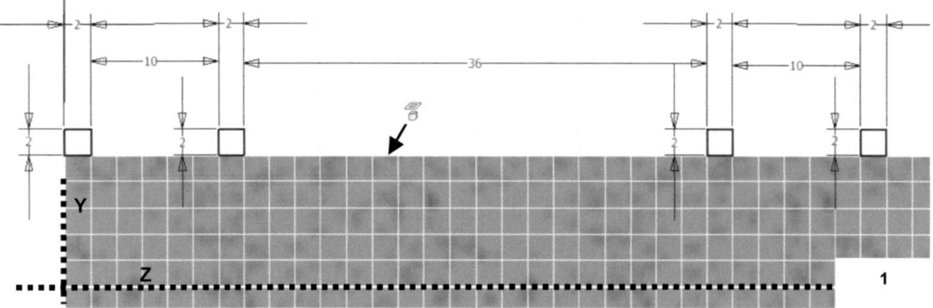

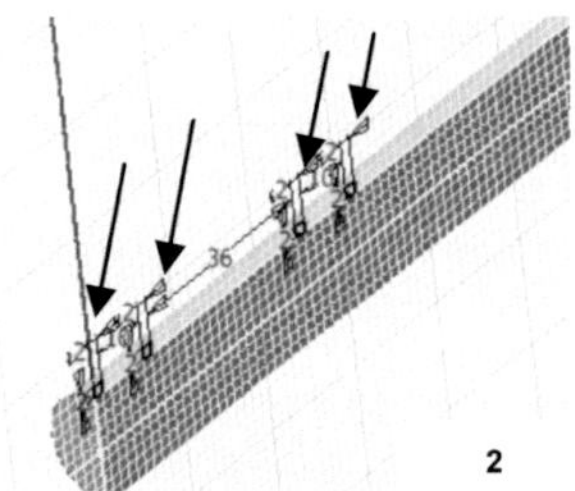

> ☑ **2D Sketch** (on Y-Z-plane)
> ⬛ **Project Geometry** (X-Y-Z-axes and marked edge of the lobe)
> Slice sketch with **F7**
> Draw 4 rectangles according to sketch
> ✔ **Finish Sketch**

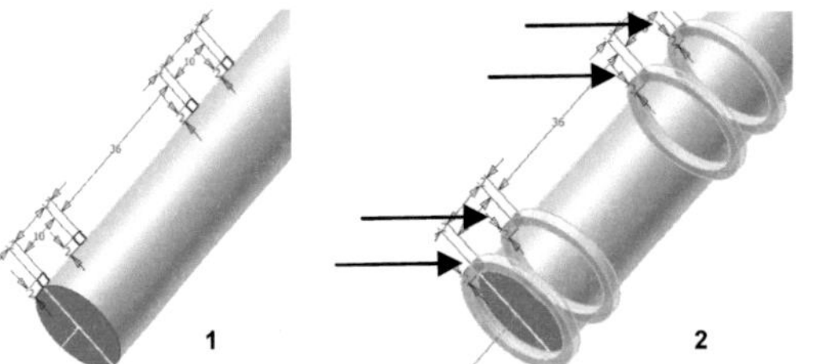

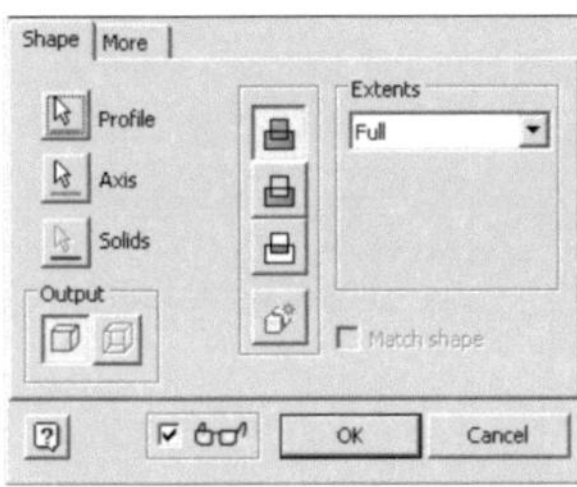

> ⬛ **Revolve**
> Profile: 4 Rectangles
> Axis: Z-axis
> Operation: Join
> Extents: Full
> ☐ OK

Via command ⬛ **Rectangular Pattern** we create the remaining 12 guides.

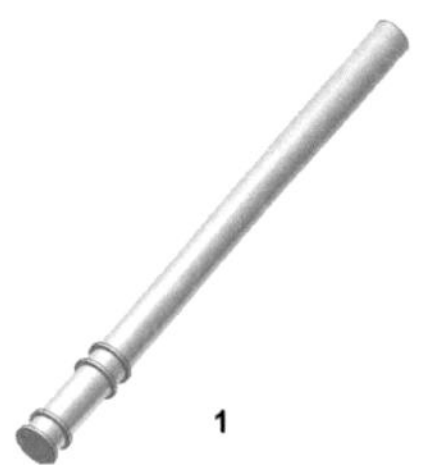

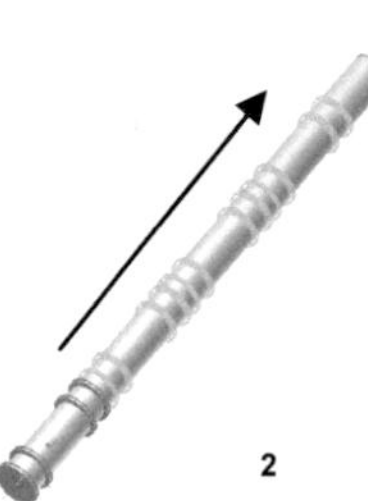

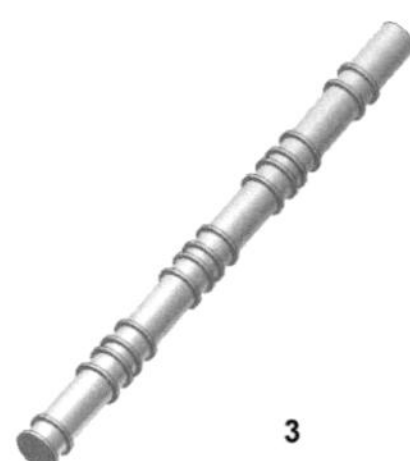

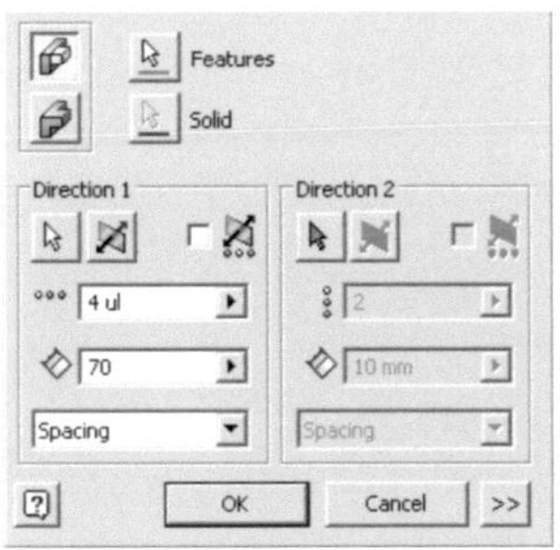

> ⊟ **Rectangular Pattern**
> Elements: Created revolve
> Direction: Z-axis
> Quantity: 4
> Distance: 70
> [OK]

6.13.4 Create cams

For the first two cams we need 2 new ▢ **Planes**. Create those with a distance of 22 and 42 to the X-Y-plane. Then use the command ⊟ **Rectangular Pattern**, to create 3 further plane pairs in a distance of 70.

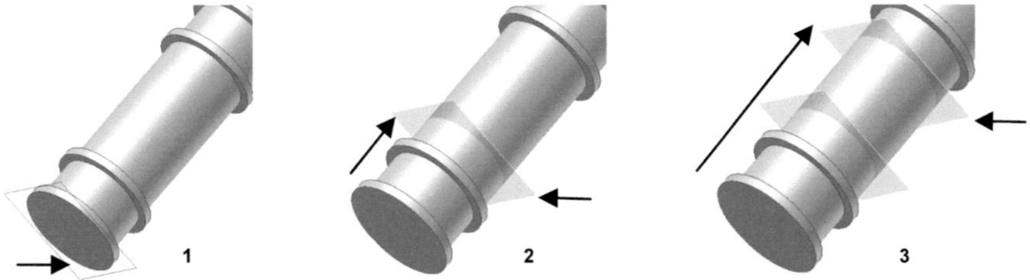

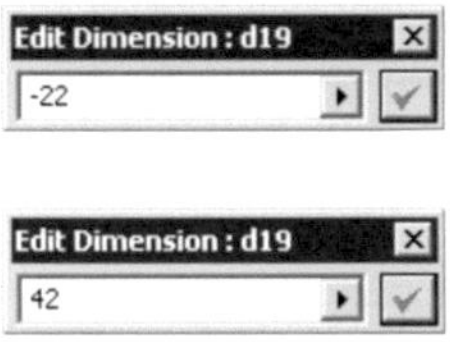

> **Plane** (create first plane)
> Distance: 22 (to X-Y-plane)
> ✓
>
> **Plane** (create second plane)
> Distance: 42 (to X-Y-plane)
> ✓

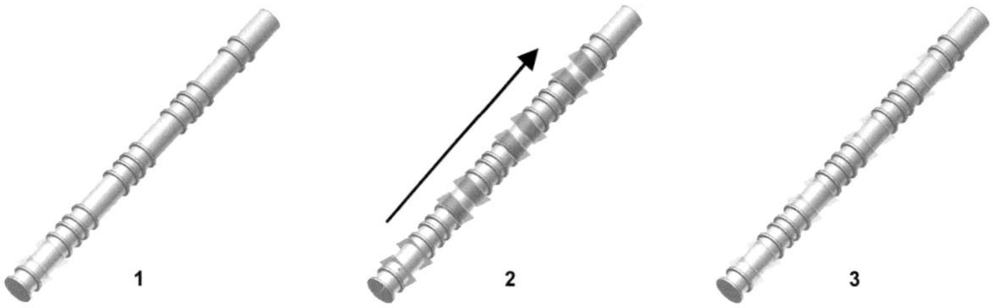

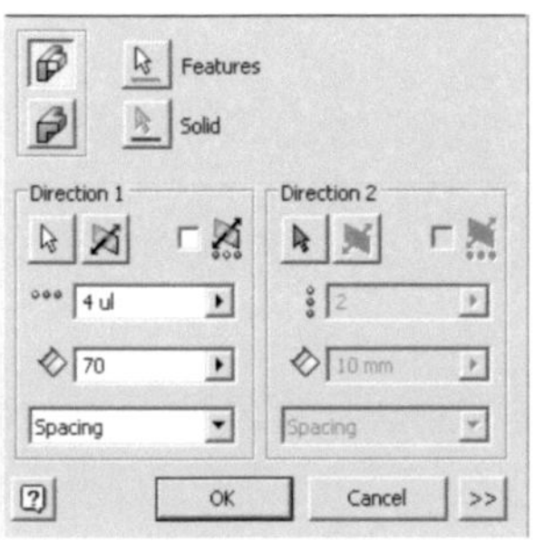

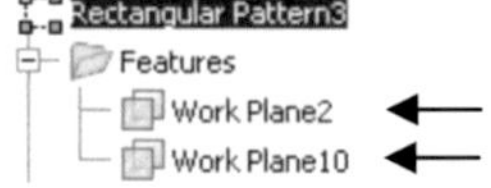

- > ⌗ **Rectangular Pattern**
- > Elements: Both created planes
- > Direction: Z-axis (picture 2)
- > Quantity: 4
- > Distance: 70
- > [OK]

On the first plane create a new ✎ **2D Sketch**, slice it with **F7**, project the X-Y-Z-axes and draw the 2 circles (D1 = 24, D2 = 10), which you connect with two tangents.

Then extrude the area symmetrically by 8.

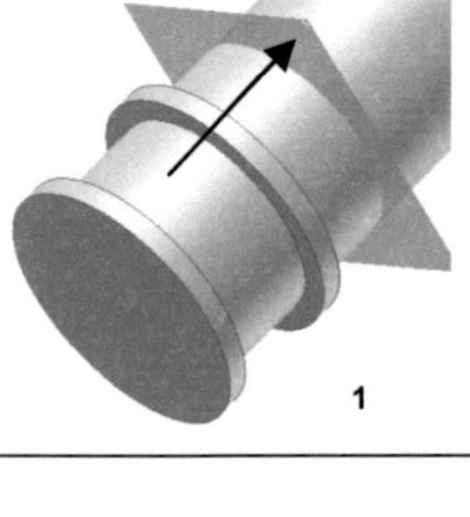

1

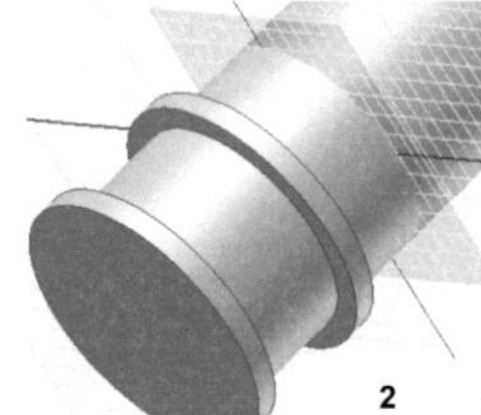

2

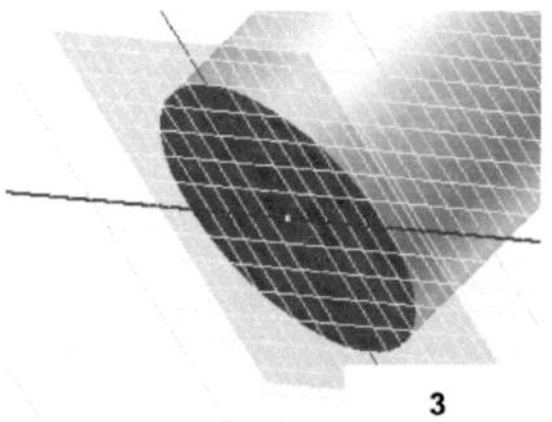

3

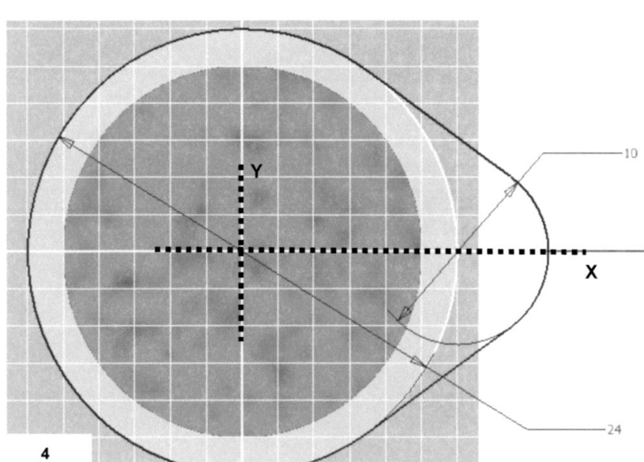

- > ✎ **2D Sketch** (on shown plane)
- > Slice with **F7**
- > ⌗ **Project Geometry** (X-Y-Z-axes)
- > Draw circles (D1= 10, D2= 24) and connect with 2 tangents
- > ✔ **Finish Sketch**

The center point of the circle with D= 10 is provided at the intersection point of the x-axis and the cirlce D= 24.

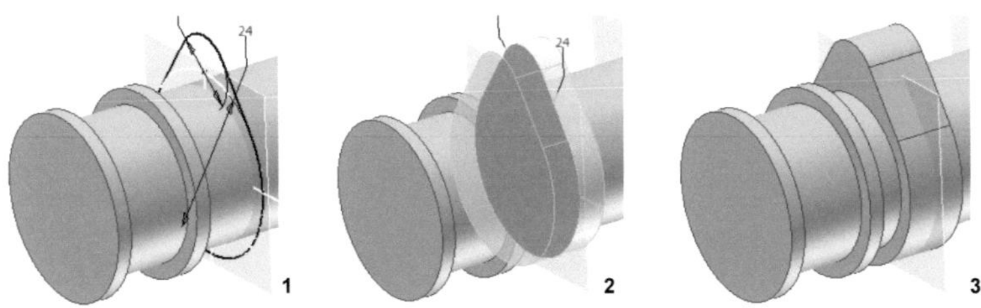

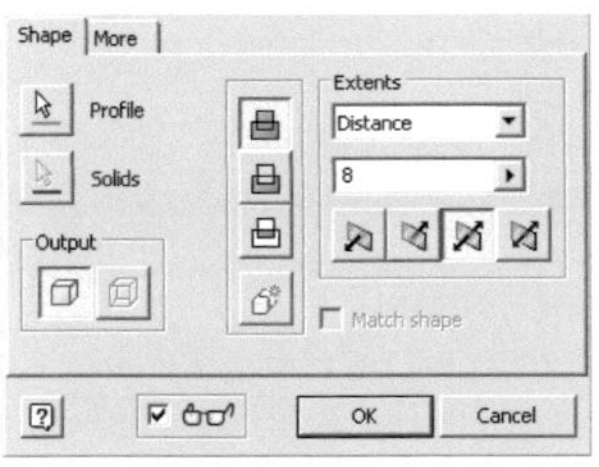

> 🗌 **Extrude**
> Profile: Area
> Operation: Join
> Extents: Distance 8
> Direction: Symmetrical
> ☐ OK

The second cam we create with **Copy** and **Paste** of elements. Mark the first cam within the model tree and copy it (**Strg+C**). Then paste it (**Strg+V**). The shown window opens. As the intake and exhaust valve have different engine valve timings the second cam has to be inserted at 90° counterclockwise. Therefore you first choose the second plane and enter an angle of 90° in the **Paste Features** window.

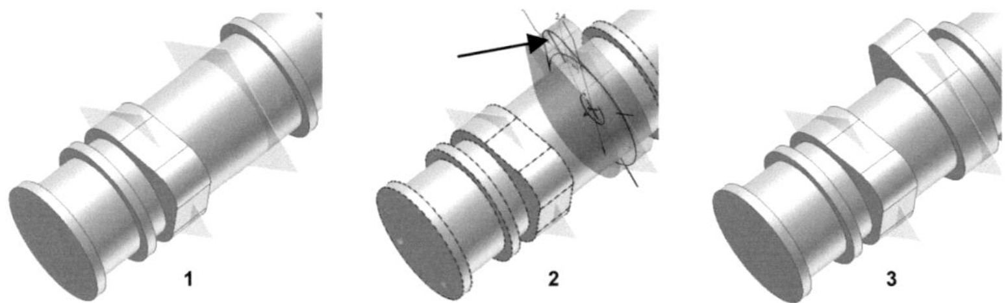

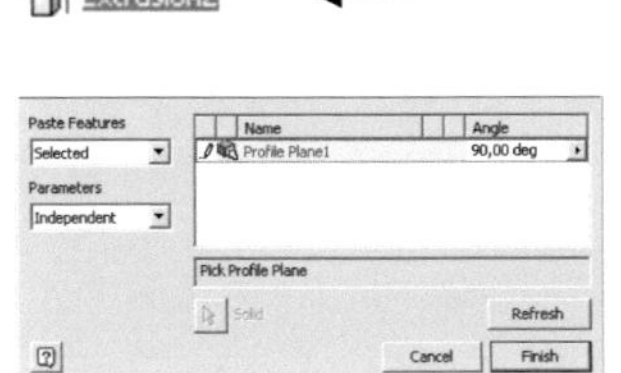

> Mark <u>first</u> cam within the model tree
> **Strg+C** (copy extrude)
> **Strg+V** (paste extrude)
> 2. Created plane
> Parameters: Independent
> Angle: 90°
> ☐ OK

Tip: Before the copy can get pasted into the drawing area you have to click **LMC** on an area within the model tree or in the drawing (not a part).

The intake and exhaust cams of the first cylinder are created. Now you can create the cams of the cylinders 2-4. Repeat the **Copy** and *Paste* twice. Mind the quantity of the elements having to be copied. As each cylinder has different valve timing the stated angle has to be taken into account as well!

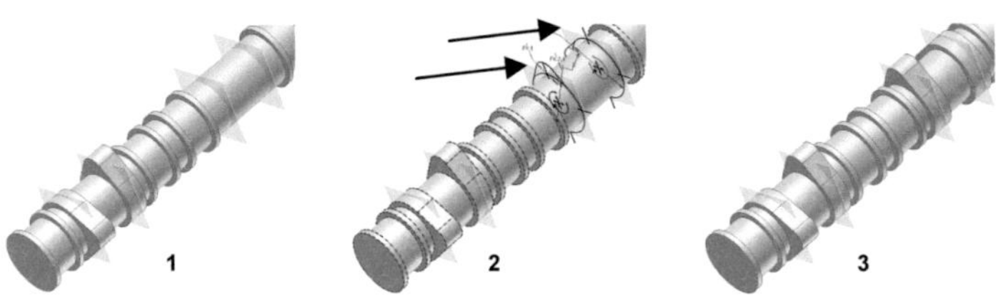

> ➢ Mark <u>both</u> cams within the model tree
> ➢ *Strg+C* (copy extrude)
> ➢ *Strg+V* (paste extrude)
> ➢ 3. Created plane
> ➢ Parameters: Independent
> ➢ Angle: 90°
> ➢ 4. Created plane
> ➢ Parameters: Dependent
> ➢ Angle: 180°
> ➢ OK

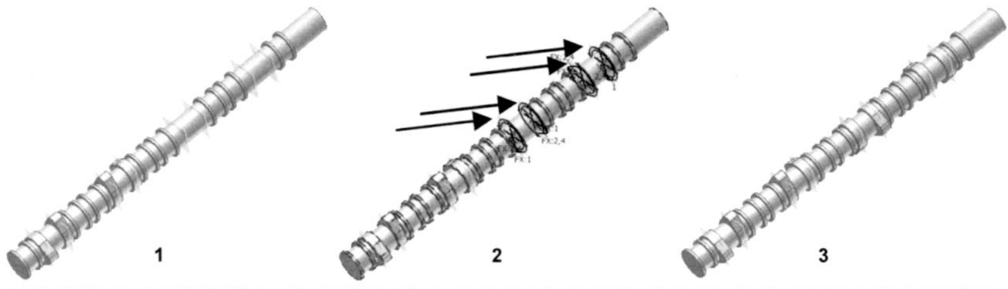

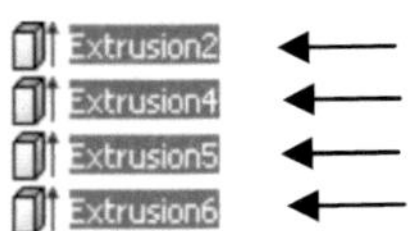

> ➢ Mark the <u>four</u> cams within the model tree
> ➢ *Strg+C* (copy extrude)
> ➢ *Strg+V* (paste extrude)
> ➢ 5. Created plane
> ➢ Parameters: Independent
> ➢ Angle: 270°
> ➢ 6. Created plane
> ➢ Parameters: Independent
> ➢ Angle: 0°
> ➢ 7. Created plane
> ➢ Parameters: Independent

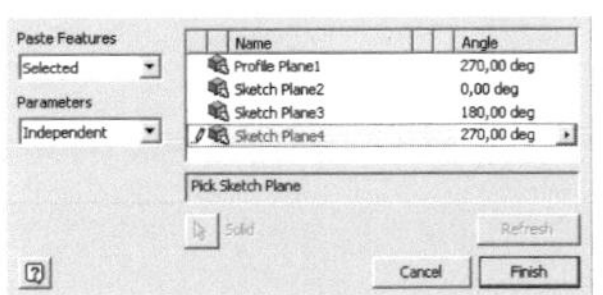

> Angle: 180°
> 8. Created plane
> Parameters: Independent
> Angle: 270°
> [OK]

Then remove the visibility from the 8 created planes.

6.13.5   Fitting key notches of the shaft's ends

To tightly connect the camshaft to the pulley we need a fitting key notch. First create a plane with the value increased by 6 to the X-Z-plane, then create a new ✍ **2D Sketch** on it, project the X-Y-Z-axes as well as the existing shaft's end (marked page picture 3) and draw the shown fitting key notch. Then extrude the volume shape.

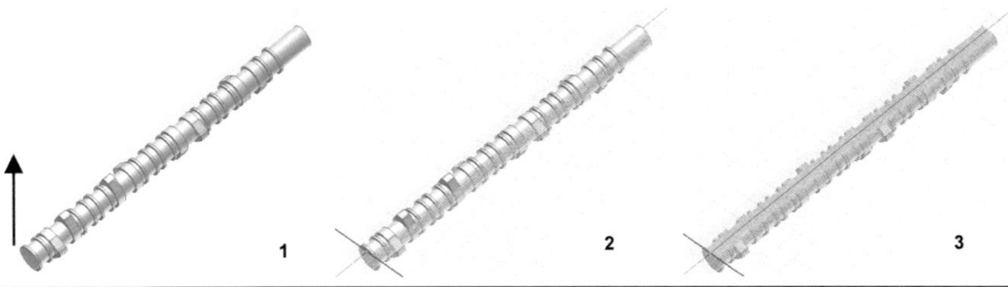

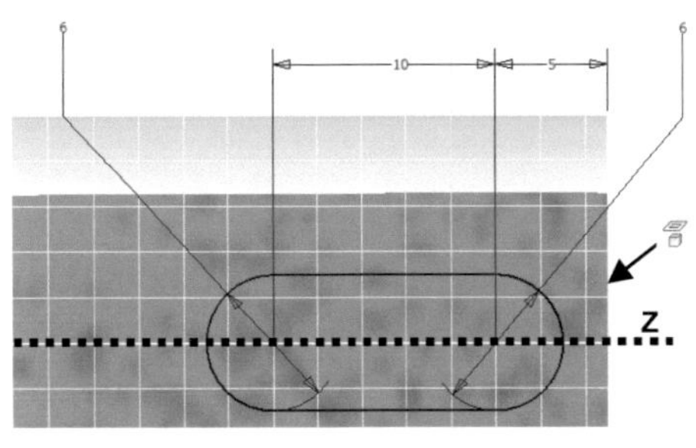

> ⬜ **Plane** (create first plane)
> Distance: 6 (to X-Z-plane, direction showed in picture 1)
> ✔

> ✍ **2D Sketch** (on new plane)
> Slice sketch with **F7**
> ⬆ **Project Geometry** (X-Y-Z-axes and right face of the camshaft)
> Draw shown outline
> ✔ **Finish Sketch**

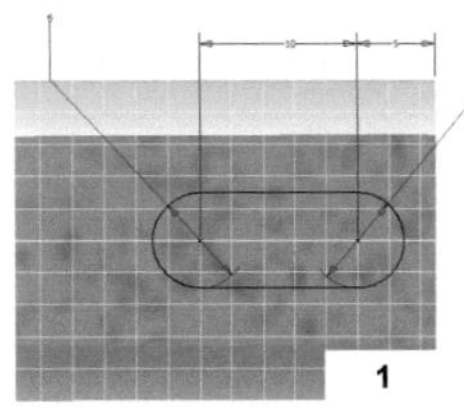

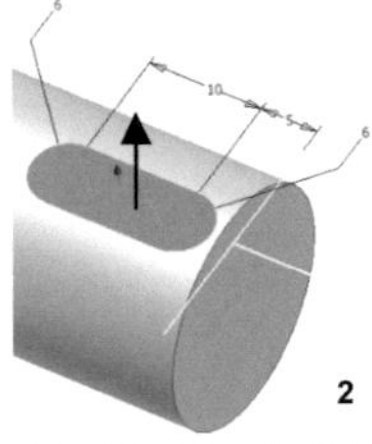

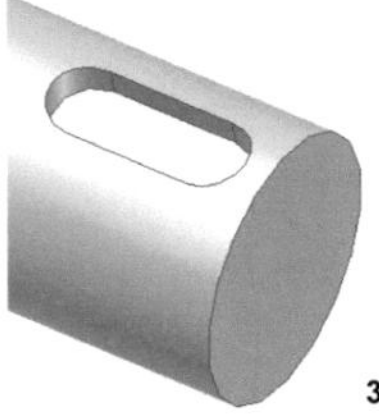

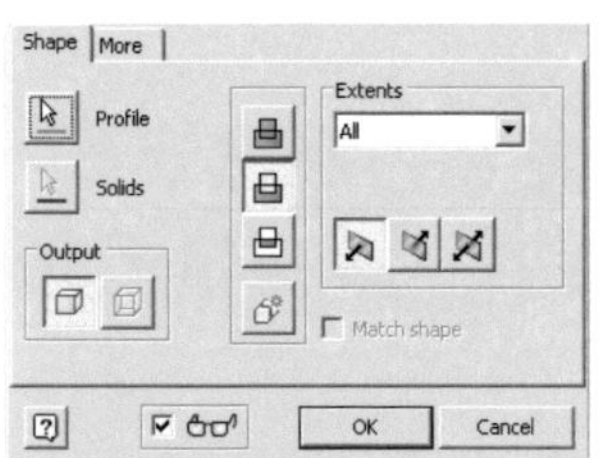

> ▸ ▯ *Extrude*
> ▸ Profile: Area
> ▸ Operation: Cut
> ▸ Extents: All
> ▸ Direction: Shown (picture 2)
> ▸ ☐ OK

6.13.6   Threaded holes on the end face areas

Finally we need a ▣ *Threaded hole* to later secure the pulley against axial drifts. For this create a new ▣ *2D Sketch* on the camshaft end face (front end under the fitting key notch), project the X-Y-Z-axes and place a ✛ *Point* in the center to then be able to execute the command ▣ *Hole*.

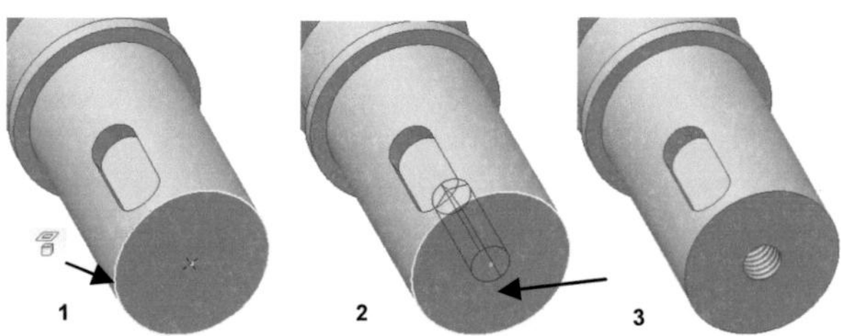

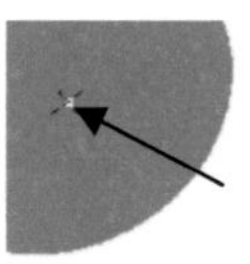

> ▸ ▨ *2D Sketch* (on marked area picture 2)
> ▸ ▧ *Project Geometry* (marked edge picture 1)
> ▸ Place ✛ *Point* (concentric to projected edge)
> ▸ ✔ *Finish Sketch*
>
> ▸ ▣ *Hole* (with thread)
> ▸ Placement: From Sketch
> ▸ Termination: Distance 15
> ▸ Thread Type: ISO Metric profile
> ▸ Size: 6 - Right hand
> ▸ Designation: M6 x 1
> ▸ Class: 6H
> ▸ Depth: Full
> ▸ ☐ OK

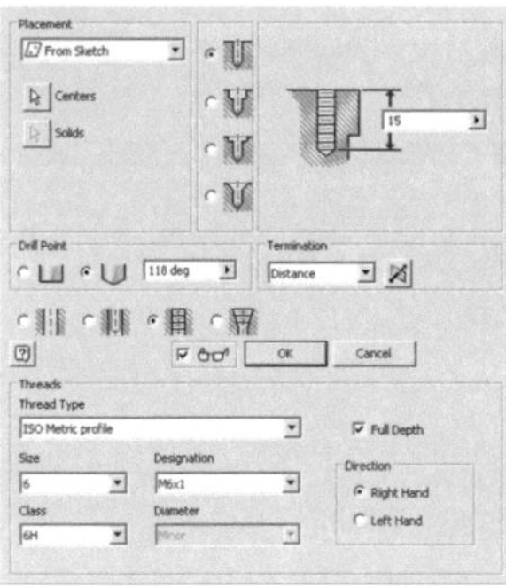

Note: If you are using *inch* units, please transform the metric ones to inch system.

camshaft.ipt now is complete. ▣ *Save* and ▣ *Close* the file.

6.14 Part CRANKSHAFT
6.14.1   Basics of the part

It's the job of the crankshaft to change the linear movement of the pistons into a rotating movement of the crankshaft. There are milled (consisting of separate parts) and forged or cast crankshafts (consisting of just one part).

The components of a crankshaft are crank webs, main bearings, connecting rod bearings, front (belt) side and drive end.

6.14.2   Crank webs

We start with the construction of the crank webs. Create a New Part and Save it as **crankshaft.ipt** in your project folder.

Project the X-Y-Z-axes and draw the shown outline. For the first 4 crank webs we need 4 **Planes**. The first plane is the X-Y-plane; three more have to be created parallel to the X-Y-plane with distances of 40, 70 and 110.

Then we extrude the sketch of the first sketch symmetrically by 10, create a new **2D Sketch** on the first newly created plane (distance 40) and project the just created outline edges onto it.

Also **extrude** these symmetrically by 10.

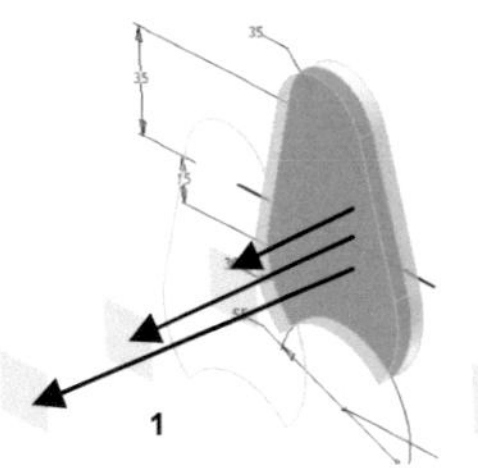
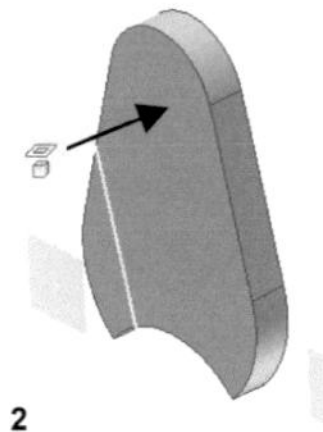
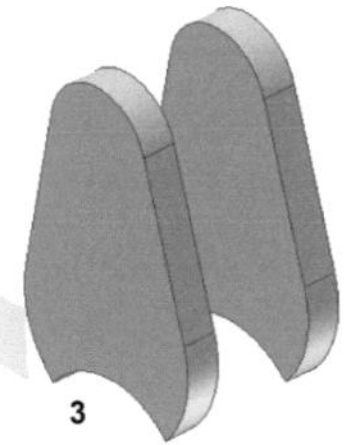

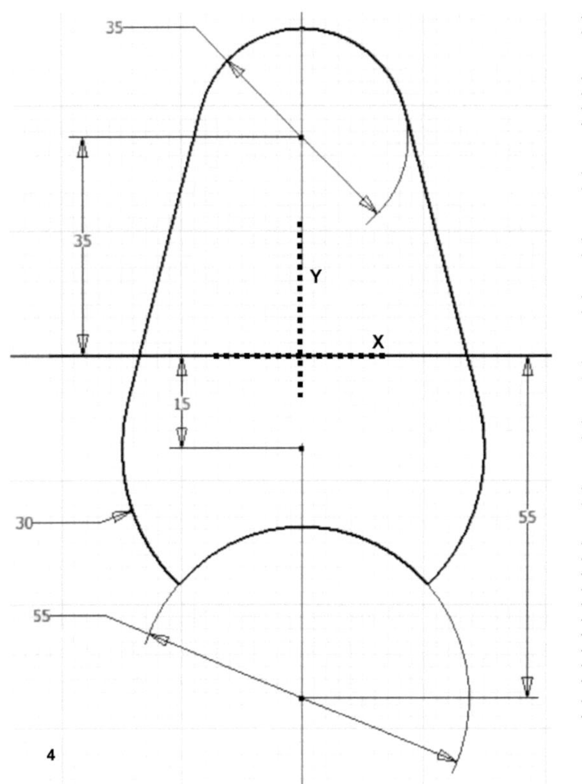

> ➢ 🖈 **Project Geometry** (X-Y-Z-axes)
> ➢ Draw shown geometry
> ➢ ✔ **Finish Sketch**
>
> ➢ 🗗 Create **1. Plane** with a distance of 40 to the X-Y-plane
> ➢ ✔
>
> ➢ 🗗 Create **2. Plane** with a distance of 70 to the X-Y-plane
> ➢ ✔
>
> ➢ 🗗 Create **3. Plane** with a distance of 110 to the X-Y-plane
> ➢ ✔
>
> ➢ 🗖 **Extrude**
> ➢ Profile: Area
> ➢ Operation: Join
> ➢ Extents: Distance 10
> ➢ Direction: Symmetrical
> ➢ [OK]

🗖 **Extrude** the newly created outline also by 10.

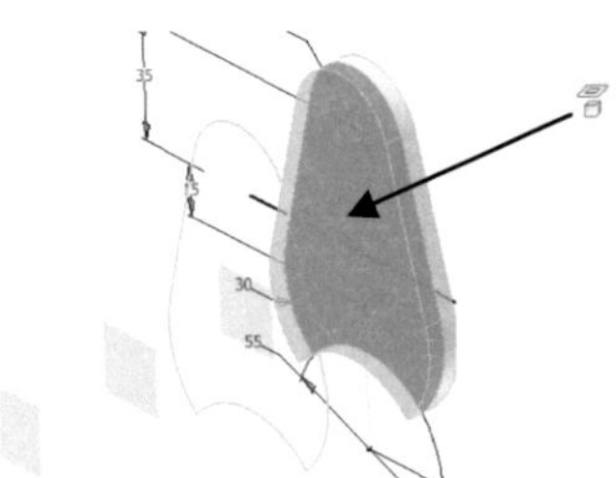

The first pair of crank webs is finished. The second pair we create by **Copy** and **Paste** with a rotation of 180°.

This time choose in the field **Paste Features** the plane with a distance of 70 (3. plane). The fourth crank web now automatically should locate itself on the fourth plane.

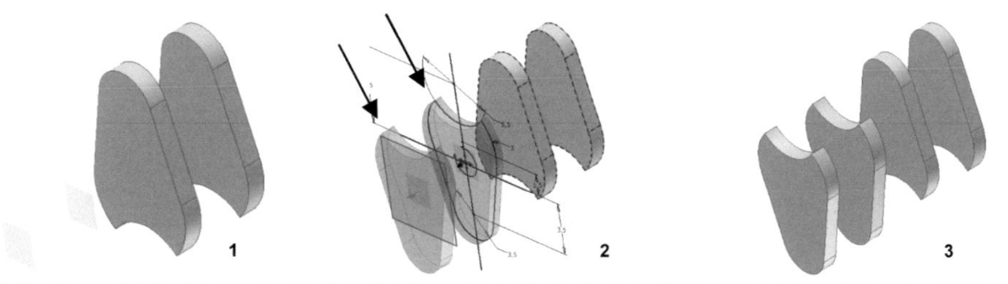

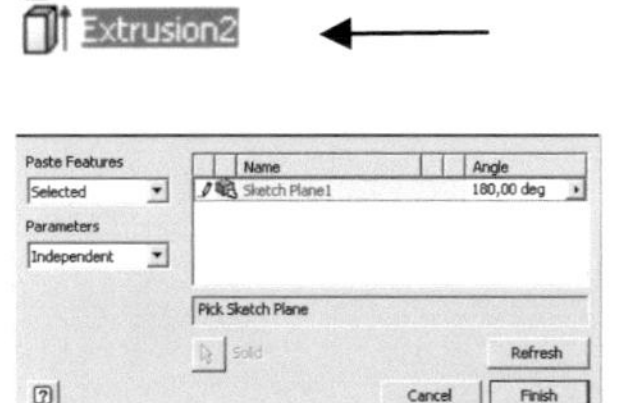

> ➤ Mark <u>both</u> crank webs in the model tree
> ➤ **Strg+C** (copy element)
> ➤ **Strg+V** (paste element)
> ➤ Plane with a distance of 70
> ➤ Parameters: Independent
> ➤ Angle: 180°
> ➤   OK

6.14.3   *Connecting rod bearings*

Create a new **2D Sketch** on the Y-Z-plane, project the X-Y-Z-axes and the marked edges of the crank webs and draw the two shown outlines.

Then execute the 2 revolves and create the two volume shapes.

Tip: Both of the outlines shown in picture 1 are geometrically identical, just mirrored by 180° (a section of the lower outline is shown in picture 3).

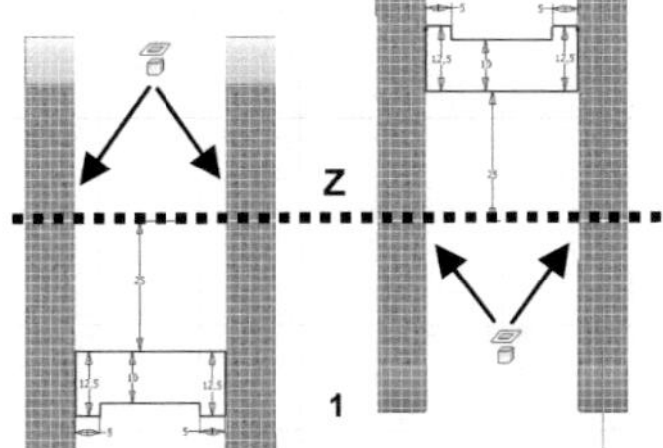

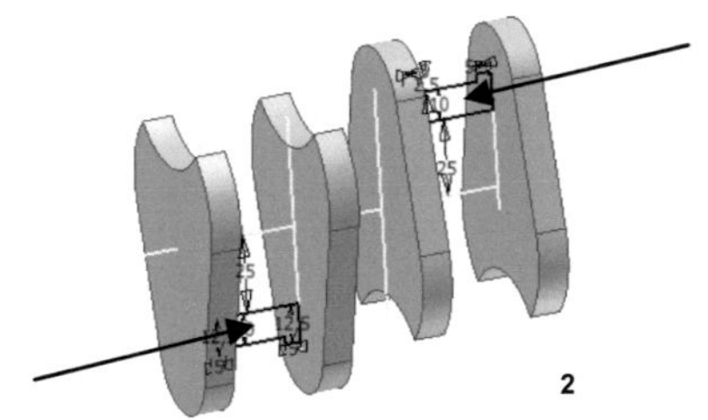

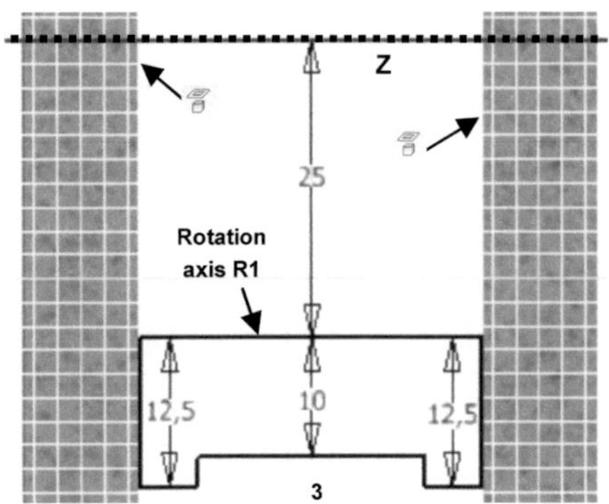

> ☑ **2D Sketch** (on Y-Z-plane)
> Slice sketch with **F7**
> ☞ **Project Geometry** (X-Y-Z-axes and both marked edges in picture 3)
> Draw shown outlines (picture 1+3) (picture 3 shows a detail of the left outline)
> ✔ **Finish Sketch**

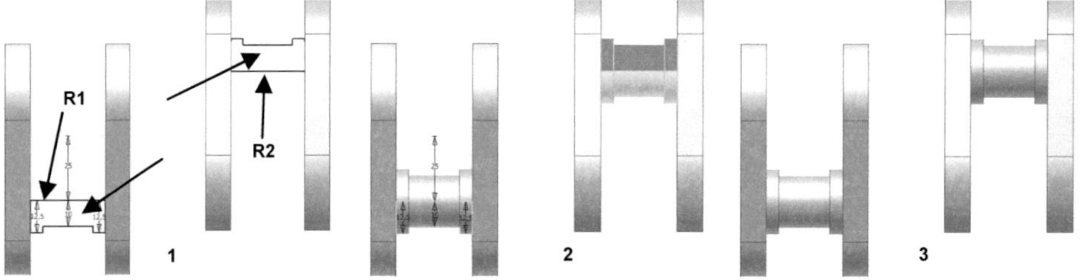

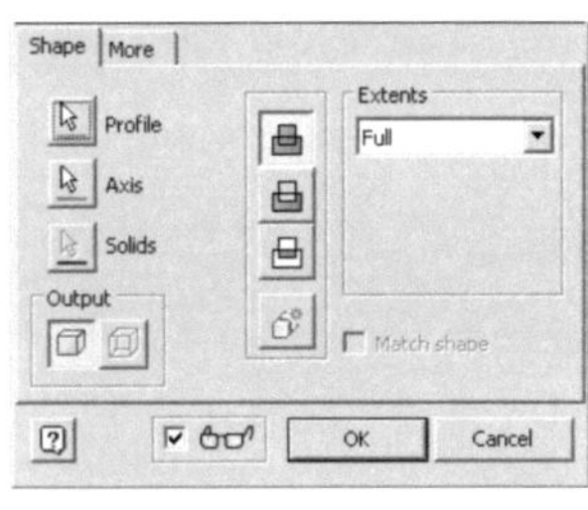

> 🛢 **Revolve**
> Profile: Area (picture 1 - left side)
> Axis: R1
> Operation: Join
> Extents: Full
> ☐ OK

> **RMC > Share Sketch**

> 🛢 **Revolve**
> Profile: Area (picture 1 - rigth side)
> Axis: R2
> Operation: Join
> Extents: Full
> ☐ OK

Tip: The respective rotation axis of the two volume shapes are tagged **R1** and **R2.**

The visibility of the sketch can now be removed again. Round off all edges (<u>besides</u> the ones marked in picture 2) with a radius of 2.

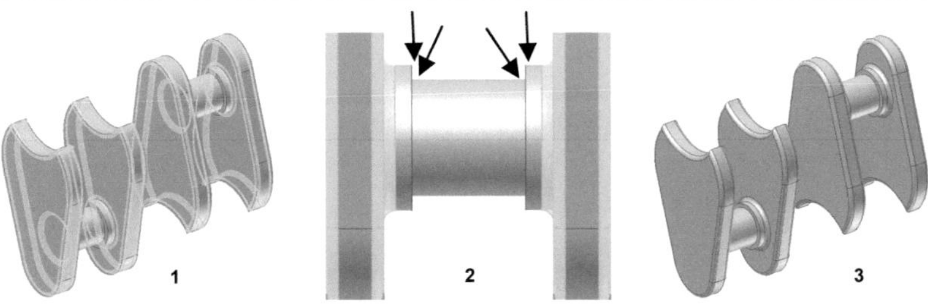

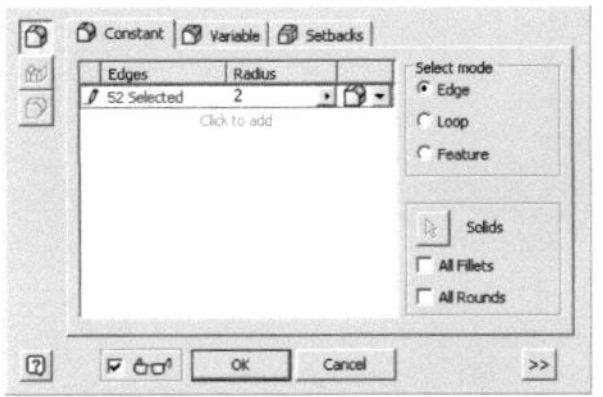

> 🖰 *Fillet*
> Edges: All edges, beside the marked edges in picture 2 (all in all 52 edges)
> Radius: 2
> [OK]

To enable mirroring of the yet created elements we need another 🖰 *Plane*, which we will create with a distance of 15 to the X-Y-plane (this time in the opposite direction as the previous 3 planes).

Then mirror all elements on this plane.

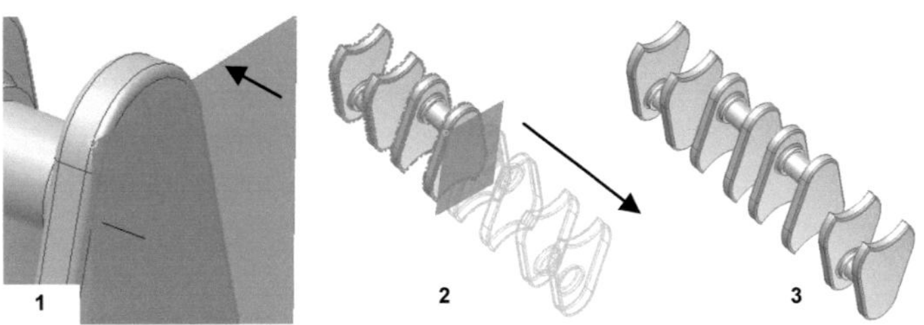

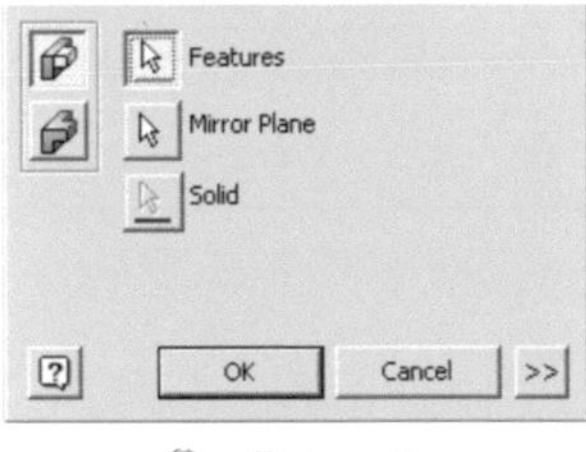

> 🖰 *Plane*
> Distance: 15 (to X-Y-plane picture 1)
> ✔

> 🖰 *Mirror*
> Elements: All yet created extrudes, rotations and roundings
> Mirror Plane: Plane (picture 2)
> [OK]

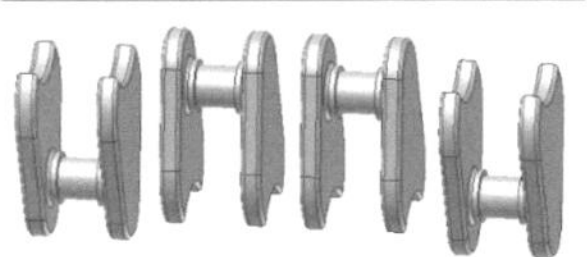

6.14.4 Main guide bearings

The 5 main guide bearings of our crankshaft enable a rotation of the crankshaft and prevent its axial drift. Create a new ✎ **2D Sketch** on Y-Z-plane, project the 3 axis and the marked body edge and then draw the shown outline.

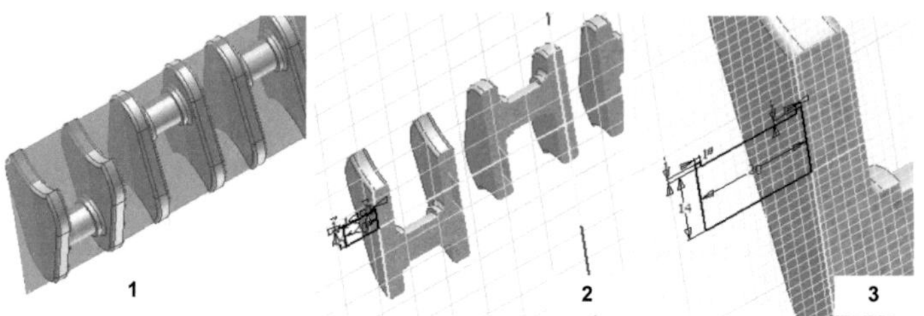

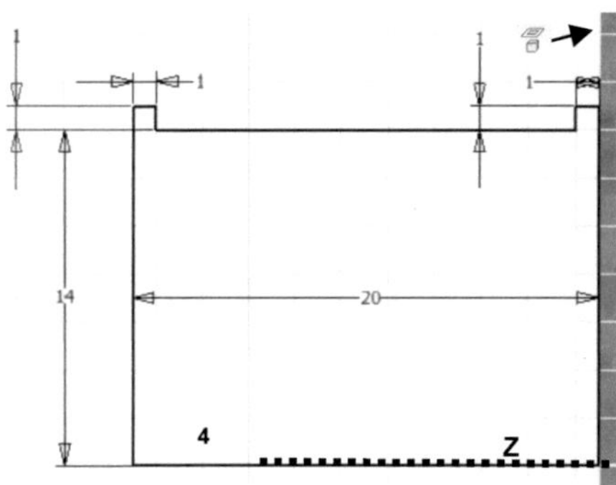

> ✎ **2D Sketch** (on Y-Z-plane)
> Slice sketch with **F7**
> ☞ **Project Geometry** (X-Y-Z-axes and marked edge)
> Draw shown outline
> ✔ **Finish Sketch**

Then rotate the area around the Z-axis and use the command ⊞ **Rectangular Pattern**, to create the remaining 4 guide bearings.

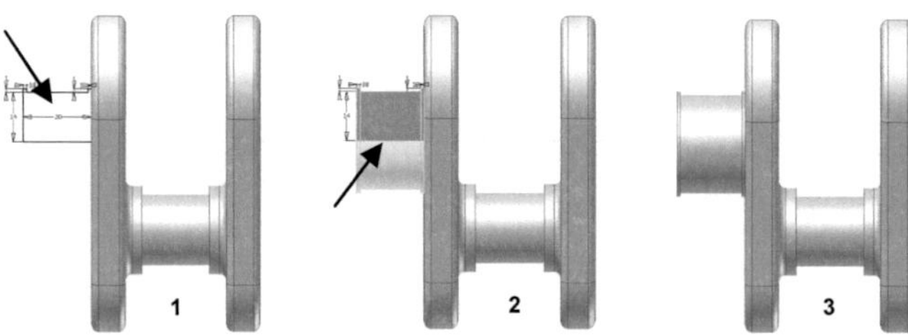

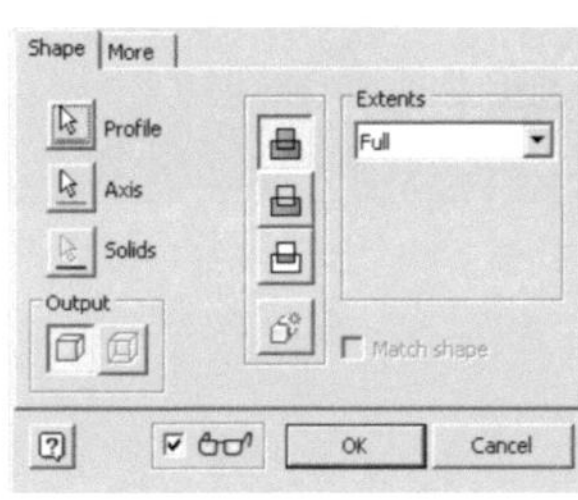

> ⊙ **Revolve**
> Profile: Marked area (picture 1)
> Axis: Z-axis
> Operation: Join
> Extents: Full
> ☐ OK

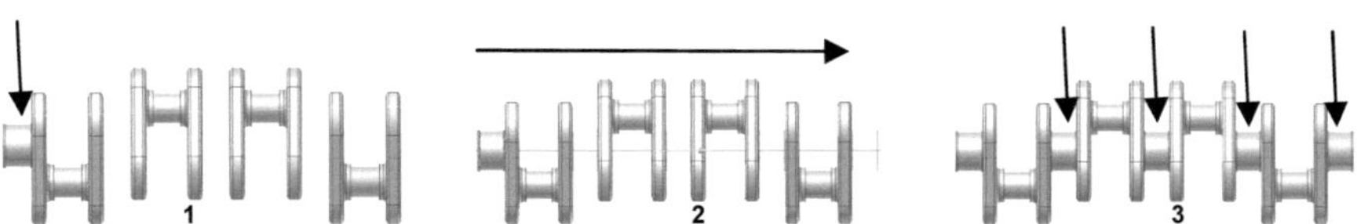

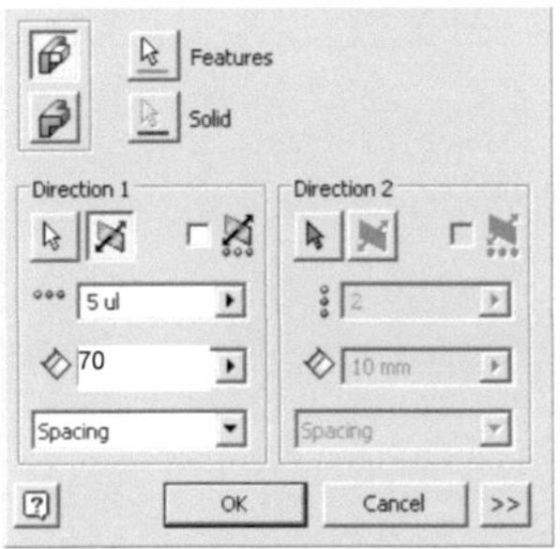

> ⠿ ***Rectangular Pattern***
> ➤ Elements: created rotary body (picture 1)
> ➤ Direction: Z-axis
> ➤ Quantity: 5
> ➤ Distance: 70
> ➤ [OK]

6.14.5 *Fitting key notch & thread hole of the belt side*

To finish we need one shaft with fitting key notch and thread hole for each, belt and drive end on both face sides of the crankshaft.

Create a new ✎ **2D Sketch** on the Y-Z-plane, project the face of the crankshaft as well as the 3 axis and then rotate the volume shape.

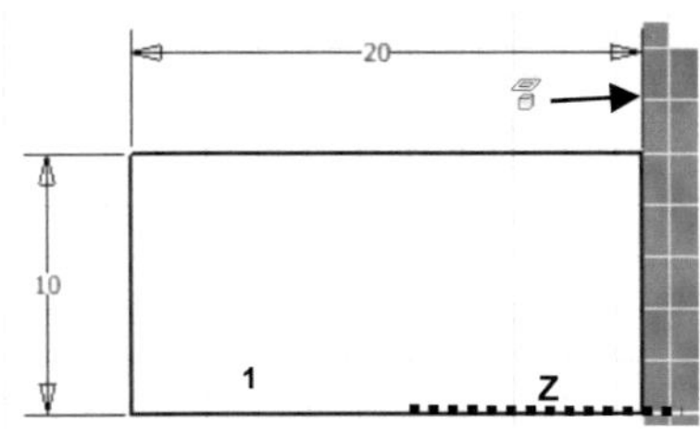
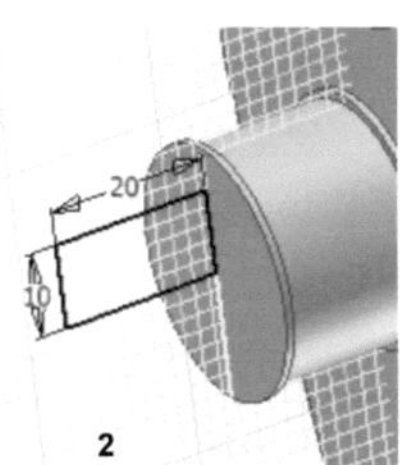

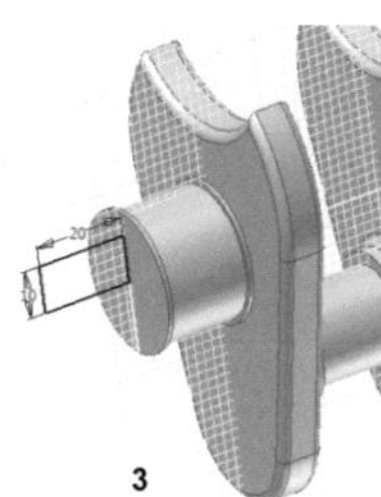

> ➤ ✎ **2D Sketch** (on Y-Z-plane)
> ➤ Slice the sketch with **F7**
> ➤ ☐ **Project Geometry** (X-Y-Z-axes and outside edges of the marked face)
> ➤ Draw shown geometry
> ➤ ✔ **Finish Sketch**

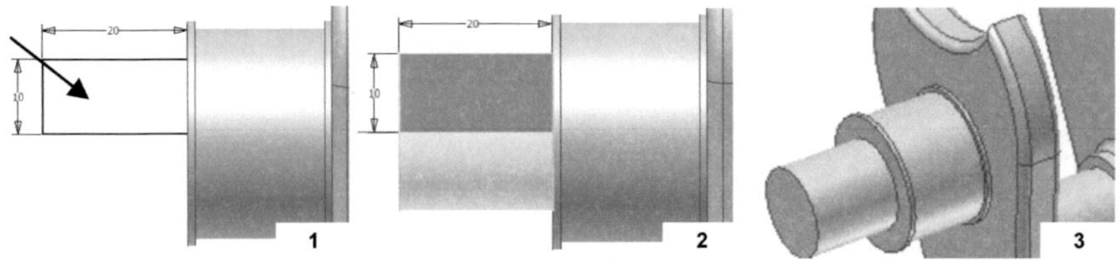

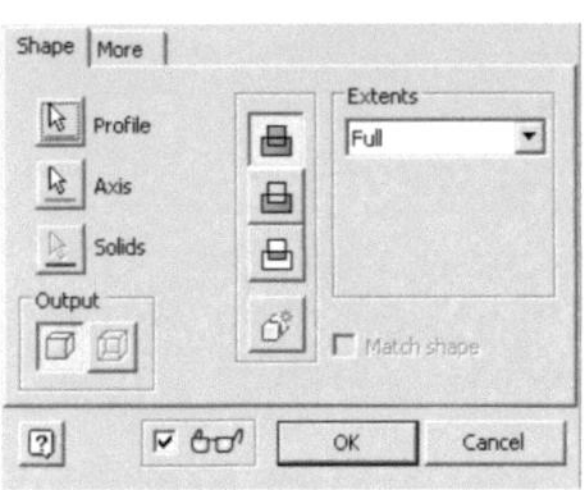

> ⊜ ***Revolve***
> ➤ Profile: Marked area (picture 1)
> ➤ Axis: Z-axis
> ➤ Operation: Join
> ➤ Extents: Full
> ➤ [OK]

For the fitting key notch we need a further ▣ ***Plane*** with a distance of 6 to the Y-Z-plane. On this you then create a new ▨ ***2D Sketch*** and draw the shown outline. With ▤ ***Extrude*** you then create the necessary material cut.

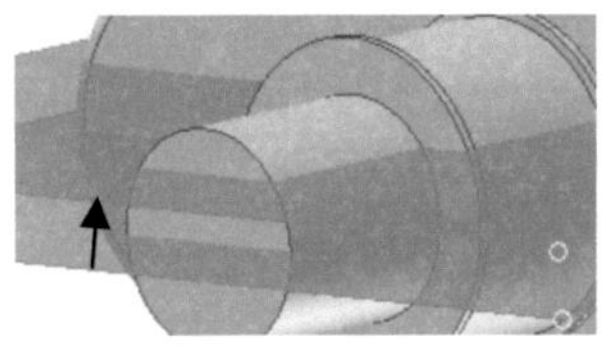

> ➤ ▣ ***Plane***
> ➤ Distance: 6 (to Y-Z-plane)
> ➤ ✔

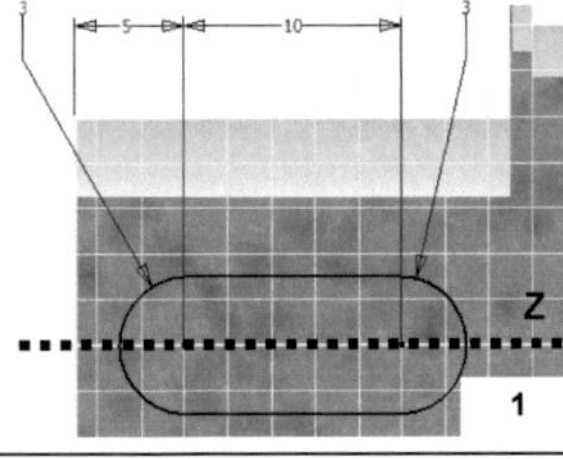

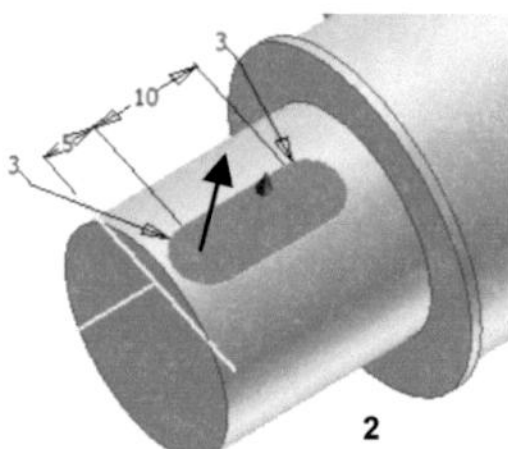

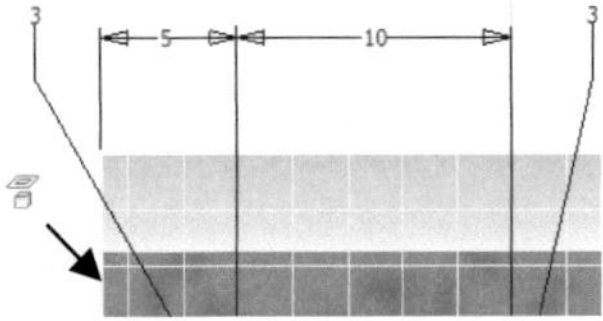

> ➤ ▨ ***2D Sketch*** (on created plane)
> ➤ Slice the sketch with ***F7***
> ➤ ▥ ***Project Geometry*** (X-Y-Z-axes and marked outside edge)
> ➤ Draw shown geometry (picture 1)
> ➤ ✔ ***Finish Sketch***

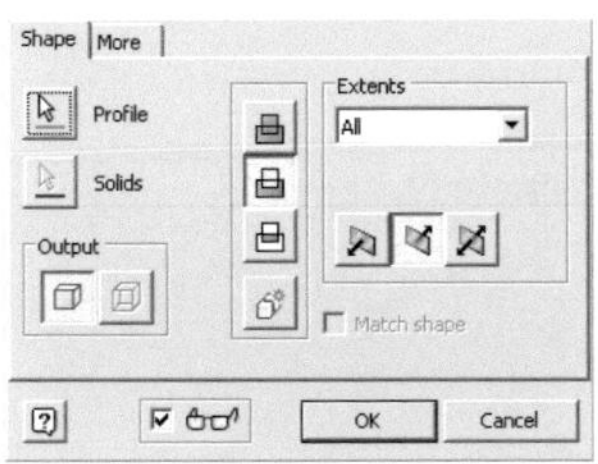

> ⬛ **Extrude**
> Profile: Area
> Operation: Cut
> Extents: All
> Direction: Shown (picture 2)
> [OK]

For the threaded hole we need a ▨ **2D Sketch** on the face of the new shaft. Place a ⊹ **Point** in the center and create the threaded hole by using the command ⊙ **Hole**.

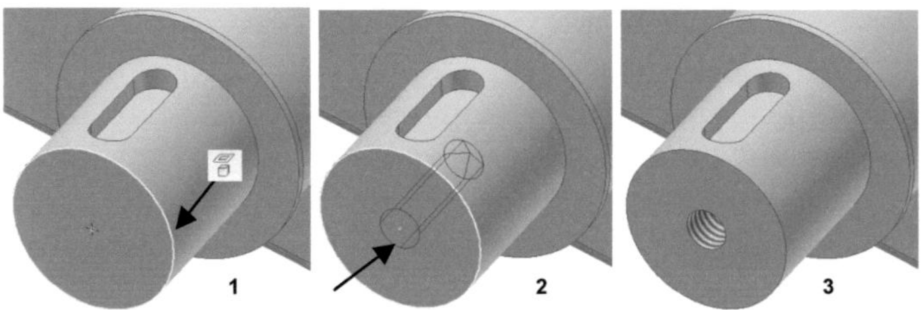

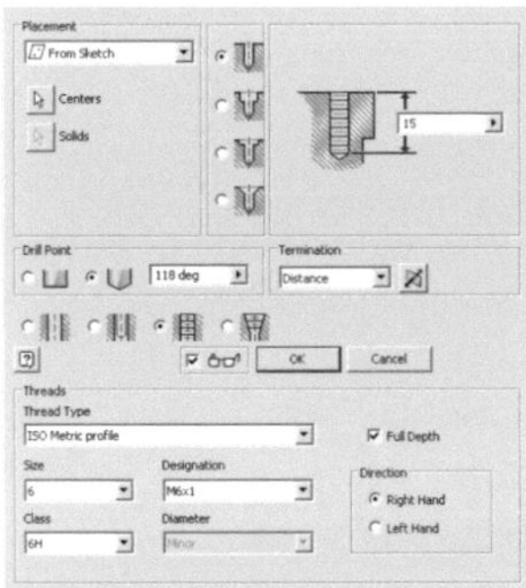

> ▨ **2D Sketch** (on face)
> ᵀ **Project Geometry** (marked outside edge, picture 1)
> Place ⊹ **Point** in the center (picture 2)
> ✔ **Finish Sketch**

> ⊙ **Hole** (with thread)
> Placement: From Sketch
> Termination: Distance 15
> Thread Type: ISO Metric profile
> Size: 6 - Right hand
> Designation: M6 x 1
> Class: 6H
> Depth: Full
> [OK]

Note: If you are using **inch** units, please transform the metric ones to inch system.

6.14.6   Mirroring of all elements for the drive end side

The drive end side is identical to the belt side. Therefore we simply have to mirror the last 3 elements on the already used plane.

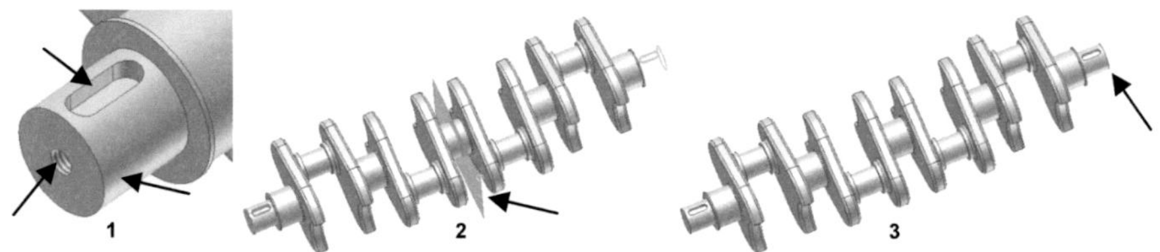

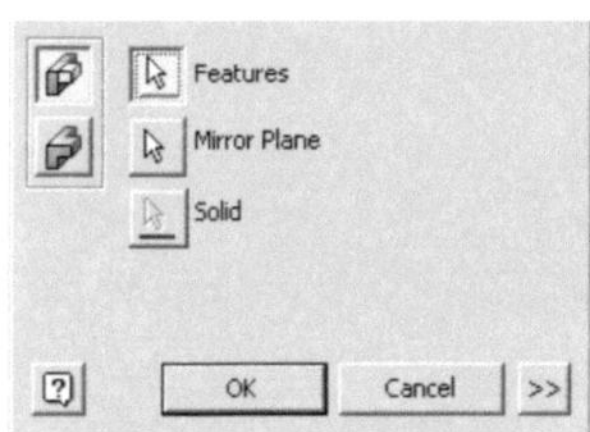

> ⫲ **Mirror**
> Elements: Rotation, extrusion and Threaded hole (picture 1)
> Mirror plane: Marked plane in picture 2
> [OK]

crankshaft.ipt now is complete. 💾 **Save** and ✖ **Close** the file.

7   ASSEMBLY GROUPS & CONSTRAINTS

7.1   Subassembly PISTON

In the construction of larger assembly groups it is quite common to use subassemblies. Here single logical components are grouped together beforehand and later inserted into the main assembly group as a subassembly.

Out main assembly **four stroke engine** consists of 5 subassemblies (assembly piston, assembly crankshaft, assembly cylinder block, assembly cylinder head, assembly camshaft) and several single components.

7.1.1   Inserting the components

For the first assembly group (assembly piston) you create a new 🗂 **New Assembly** and 💾 **Save** it as **assembly-piston.iam** in your project folder. Then you insert the following named components and a sign **Grounded** in the model tree. The first part inserted in an assembly is automatically assigned a fixation within the drawing area by the program.

The respective part has no degrees of freedom. Click the respective part in the model tree with **RMC** and remove the hook at **Grounded**.

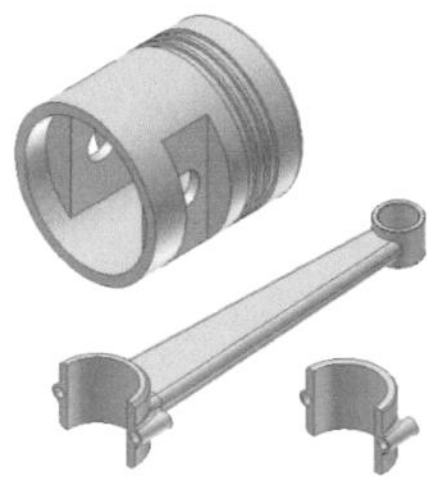

>   **Place Component**
> 1x part: piston.ipt
> 1x part: topside-piston-rod.ipt
> 1x part: bottom-side-piston-rod.ipt

Store each 1 x in the window. Storing of the components can be finished with the **ESC** key.

Tip: Part **grounded** (in the model tree) means that the part is fixed in the window and cannot be moved. The first component inserted into an assembly is automatically fixed by the program. This fixation you can remove via **RMC** > **Grounded**.

7.1.2   Setting constraints

Now the components are stored but have no affiliations yet. For this we use the command **Constrain**. It's the goal to restrict the existing freedoms of the components and by this to create defined moving constraints.

In our first example we connect the piston with the topside-piston-rod. For this we need the type **Constrain** - **Mate**.

With **Selection 1** you choose the center axis of the connecting rod end (picture 1), with **Selection 2** the center axis of the piston pin bore (picture 2) and the mode Mate.

Both components then automatically move together.

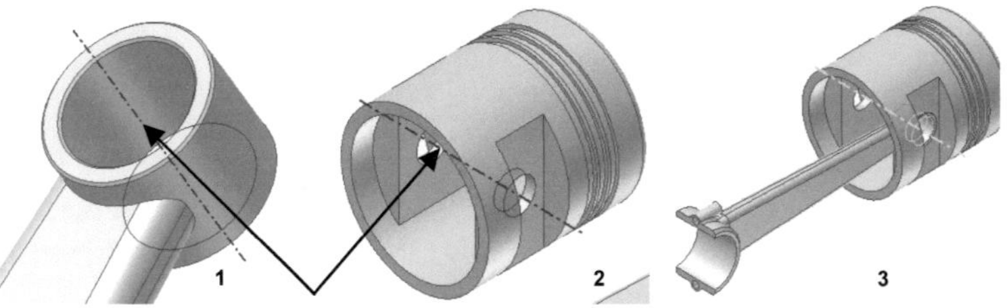

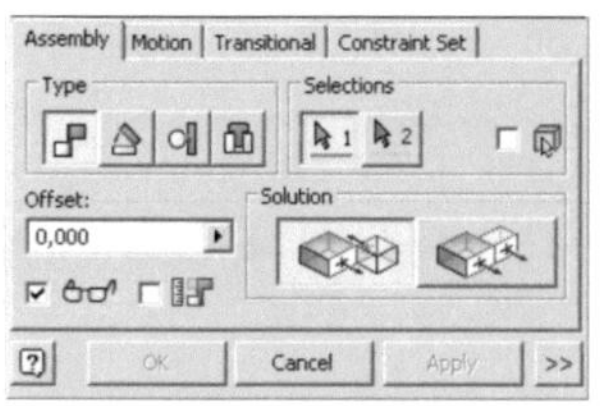

> **Constrain**
> Type: Mate
> Selections: Marked axes (picture 1+2)
> Offset: 0
> Solution: Mate
> [OK]

Tip: You can choose between the constraints **Mate**, **Angle**, **Tangent** or **Insert**. The combination of single constraints allows a precise alignment of the components between themselves and is prerequisite for a later movement of the components. With the command [Apply] **Apply** you can assign several constraints in a row.

When you click with **LMC** on the piston rod bottom side, keep the mouse key pressed and then move around in the window you see that it now is tightly secured to the chosen axis and only a movement of the components along the connecting rod end axis is possible.

We will connect the two components in a way that the piston is centered on the connecting rod end. For this we place a **Constraint** between the Y-Z-plane of the piston and the X-Y-plane of the piston rod topside.

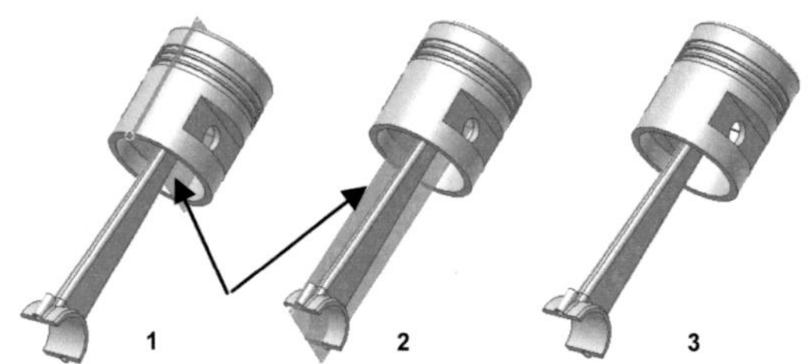
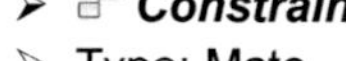

1   2   3

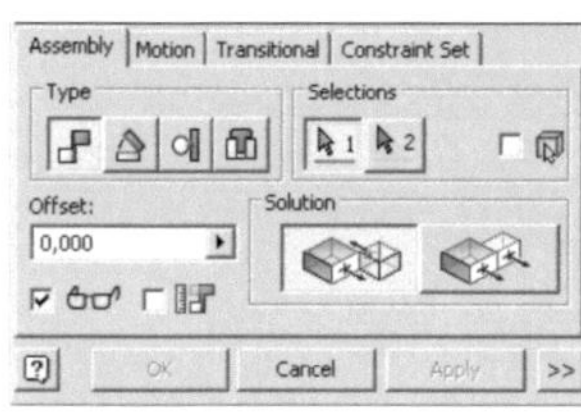

> **Constrain**
> Type: Mate
> Selections: : Y-Z-plane of piston and X-Y-plane of piston rod topside
> Offset: 0
> Solution: Mate
> [OK]

Origin
Passend:3
Fluchtend:1

When you open a part structure in the model tree you will find all constraints in the lower portion of the structure list. Here you can return to the respective editing mode and change constraints by clicking **RMC – Edit**.

In case a part has to be rotated before a constraint is assigned you can properly position it by clicking with *RMC – Component – Rotate* (or Key *G*).

A complete alignment before placing constraints is not necessary, but helps with a better general view.

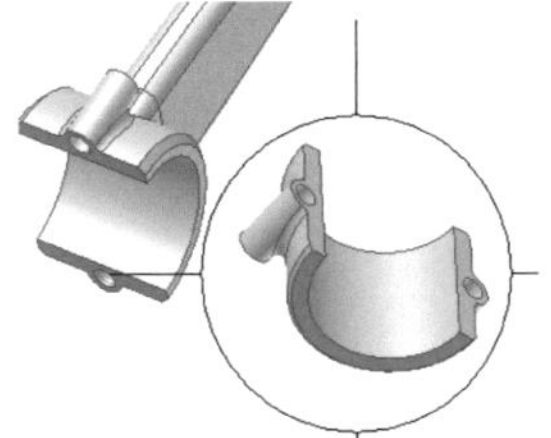

To better position the two piston rod components we now will rotate the connecting rod bottom side a little.

RMC (on *connecting rod bottom side* in the drawing window) > *Components* > *Rotate* and move to the shown position. Click *RMC – Done*. Then we assign the constraints between the two components.

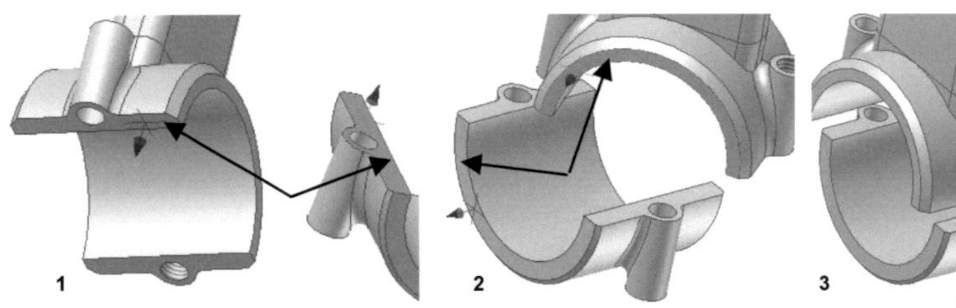

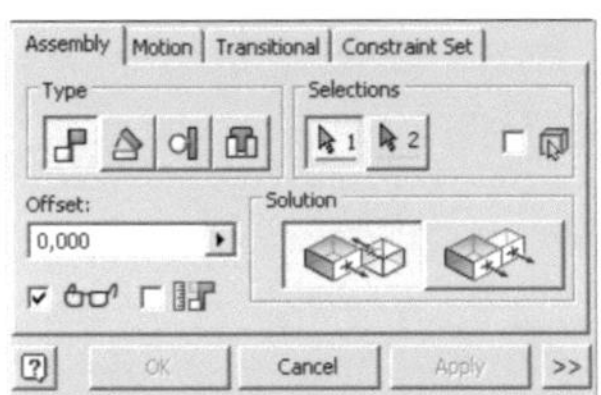

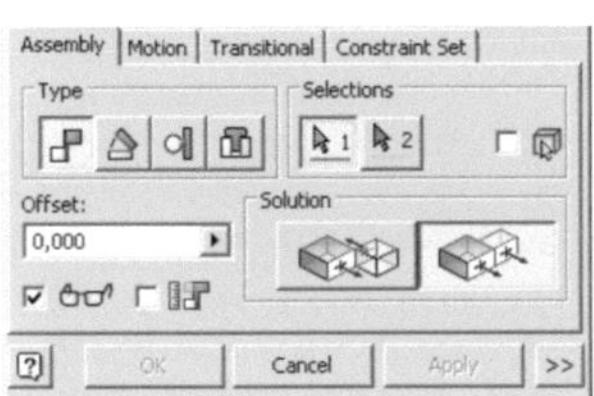

- ➢ ⬚ *Constrain*
- ➢ Type: Mate
- ➢ Selections: Marked areas picture 1
- ➢ Offset: 0
- ➢ Solution: Mate
- ➢ Apply
- ➢ Type: Mate
- ➢ Selections: Marked areas picture 2
- ➢ Offset: 0
- ➢ Solution: Flush
- ➢ Apply
- ➢ Type: Mate
- ➢ Selections: Marked axes picture 3
- ➢ Offset: 0
- ➢ Solution: Mate
- ➢ OK

Another option to create constraints is the command 🗐 **Assemble**. Following constraints can be chosen here:

- ➢ Automatic
- ➢ Mate - Mate
- ➢ Mate - Flush
- ➢ Angle - Directed
- ➢ Tangent - Outside
- ➢ Tangent - Inside
- ➢ Insert - Opposed
- ➢ Insert - Aligned
- ➢ UCS to UCS

Tip: This command can save you plenty of work. However you should carefully consider which constraints you assign, to not end up having them collide with other necessary constraints. Try out some options of this command and exercise!

7.1.3 Connecting bolts from the content center

Now the three components of the assembly group **assembly-piston** are connected. We now need two connecting bolts DIN EN ISO 4762 M3 X 20 from the 🖨 **Content Center**. The content center contains a basic selection in standard elements from the areas: sheet, elements, cable- and cable harness, profiles, ducts and conductions, other parts, connection elements and wave parts. When inserting the selected screw make sure to activate the selection 🔅 **Place more** in the automatically opened window **AutoDrop**.

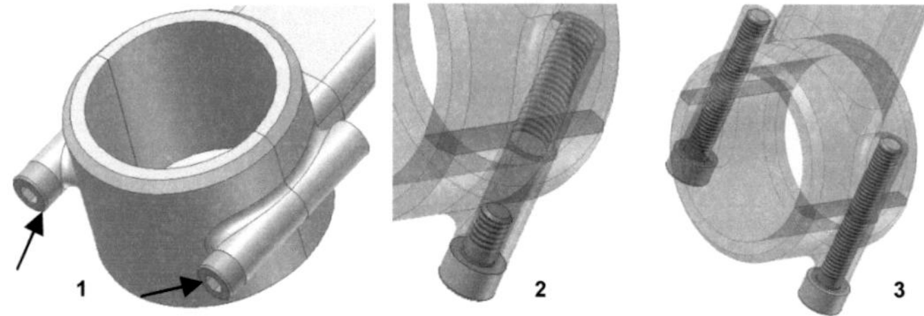

- ➢ 🖨 **Place from Content Center**
- ➢ Connecting elements – Screws – Cylinder Head – DIN ISO 4762
- ➢ Position: Threaded hole of the **piston rod bottom side**
- ➢ Activate and confirm 🔅 **Place More**
- ➢ Click with **RMC** on one of the two created screws in the

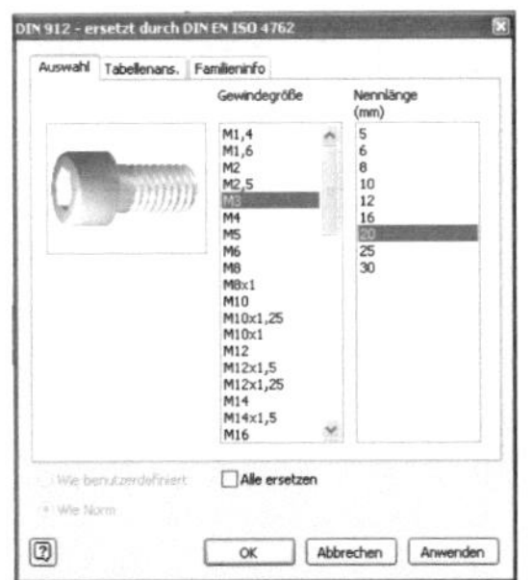

model tree to **Edit Dimensions**.

➢ Thread Size: M3
➢ Nominal Length: 20
➢ [OK]

Note: If you are using **inch** units, please transform the metric ones to inch system.

7.1.4　Creating a bolt

Now there is a connection bolt between piston and piston rod still missing. This we create directly in the assembly. Create a new part **bolt.ipt** within this assembly with the starting area marked in picture 1. You are automatically directed into the edit and sketch mode of the new part.

Here project the marked hole edge (picture 2) and end the sketch again. Then extrude the circle area to the backside of the piston and with this create a bolt length adjusted to the piston. The bolt now is created. You can leave the edit area of the bolt with the command ✦⊙ **Back**.

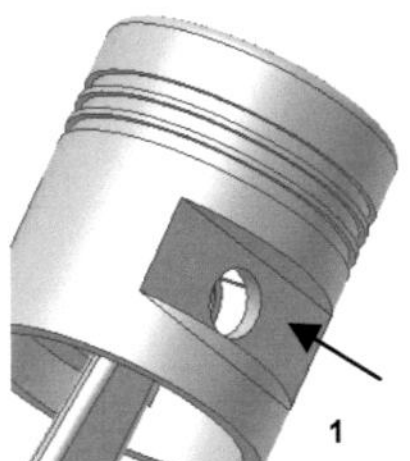

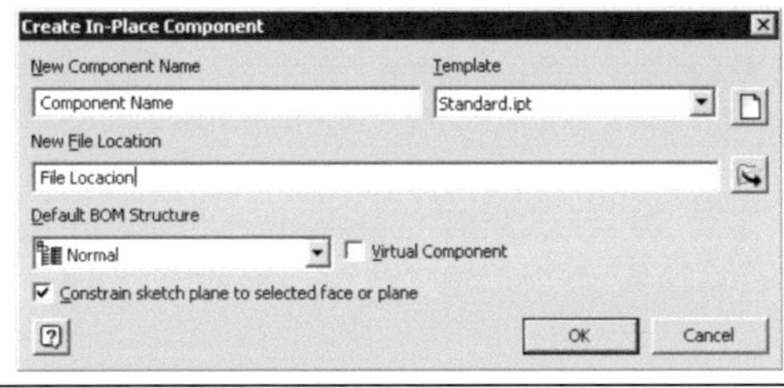

1

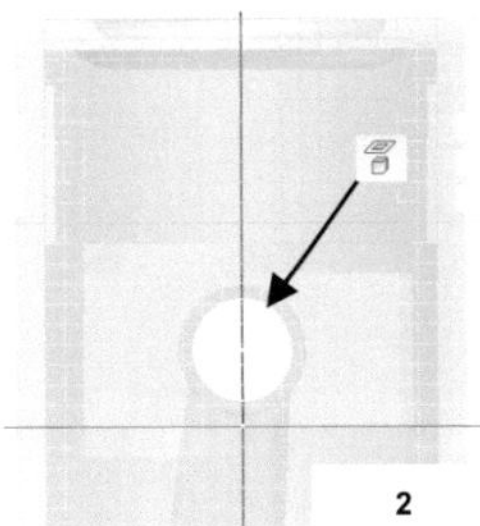

2

➢ ⬛ **Create**
➢ New Component Name: bolt
➢ Template:　Standard.ipt
➢ New File Location: Project save location
➢ Hook at: Constrain sketch plane to ...
➢ [OK]

➢ ⬛ **Project Geometry** (marked hole egde in picture 2)
➢ ✔ **Finisch Sketch**

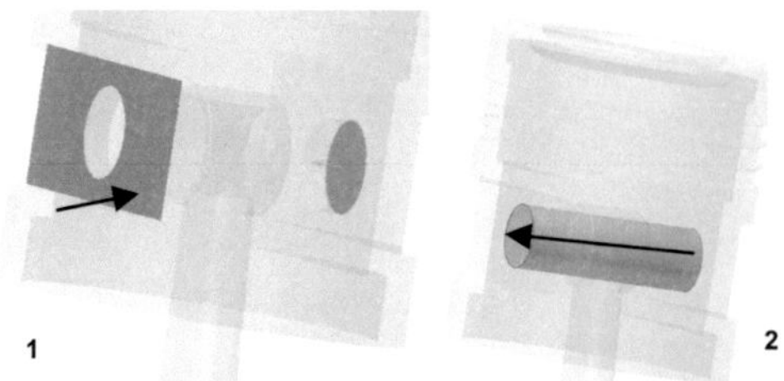

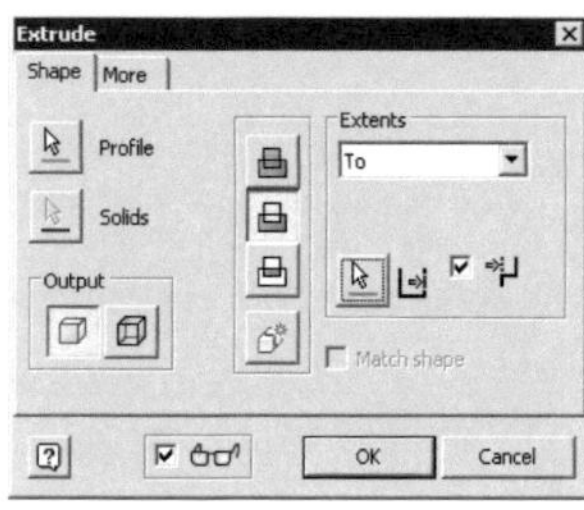

> Extrude
> Profile: Projected circle
> Operation: Join
> Extents: To
> To: Marked area picture 1
> OK

> **Back**

Tip: You can edit parts within assembly groups (the other parts of the assembly remain visible) or open the part (the part will open in a separate window). The commands are *RMC* > *Edit* or *RMC* > *Open*.

7.1.5 Allocate colors

To finally give our assembly group some color mark the respective parts in the model window and then change the color with the command *Color Override*.

The choice of color is up to you. Only color suggestions are available. The now allocated colors can be changed at any given time.

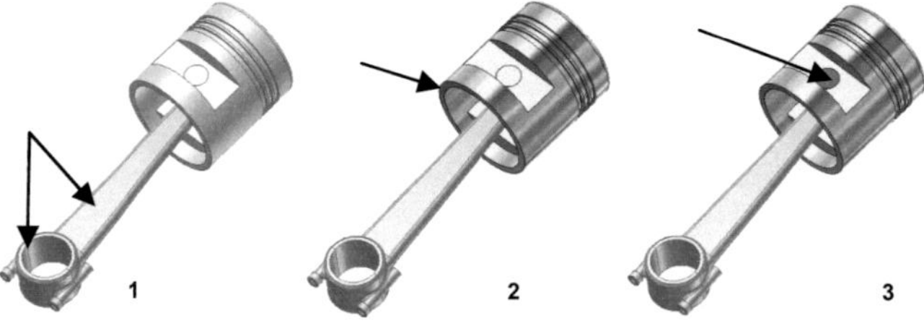

> Mark *topside piston rod* and *bottom side piston rod* in the model tree
> *Farbe* **Color Override**
> Color: e.g. metal (brass)

> Mark *piston* in the model tree
> *Farbe* **Color Override**
> Color: e.g. chrome (blue)

> Mark *bolt* in the model tree
> *Farbe* **Color Override**
> Color: e.g. red

assembly-piston.iam now is complete. 🖫 *Save* and ✖ *Close* the file.

7.2 Subassembly CRANKSHAFT

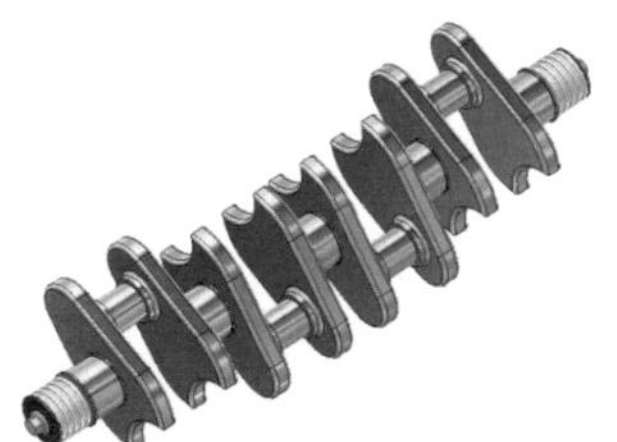

Our assembly group crankshaft will consist of the components crankshaft, pulley, fitting key, securing coupling and hexagon screw.

The fitting keys and hexagon screws we will load from the content center.

7.2.1 Inserting the components

For this assembly we create a 🗗 *New Assembly* and 🖫 *Save* it in the project folder as **assembly-crankshaft.iam**.

Insert the shown three components (1 x crankshaft, 2 x crankshaft- pulley) and then load two fitting keys DIN 6885 A 6 x 6 x 16 from the 🖳 *Content Center*.

Then you constrain the fitting keys in the fitting key notches of the two shaft ends. Make sure to mount both fitting key sides and to remove the hook from Grounded (*RMC* > *Grounded*) from the components, which might remain after inserting the parts.

> 🖺 *Place Component*
> 1x part: crankshaft.ipt
> 2x part: crankshaft-pulley.ipt

7.2.2 Fitting keys from the content center

DIN 6885 A

GB/T 1096-2003 - A

ISO 2491 C

JIS B 1301 Feder

> 🖶 **Place from Content Center**
> Shaft Parts – Fitting Keys – Fitting Keys Rectangular – Rounded – DIN 6885 A
> Wave Diameter: 17-22
> Width x Height: 6 x 6
> Nominal Length: 16
> [OK]

Note: If you are using *inch* units, please transform the metric ones to inch system.

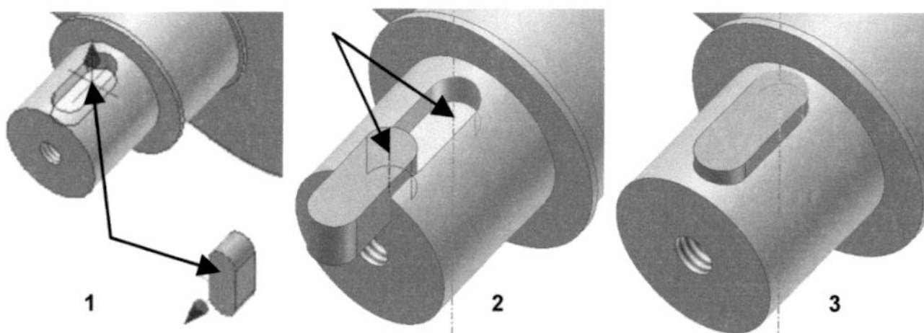

> 🔧 **Constrain**
> Type: Mate
> Selections: Marked faces (picture 1)
> Offset: 0
> Solution: Mate
> [Apply]
> Type: Mate
> Selections: Marked axes (picture 2)
> Offset: 0
> Solution: Mate
> [Apply]
> Type: Mate
> Selections: Axes on other fitting key side (picture 3)
> Offset: 0
> Solution: Mate
> [OK]

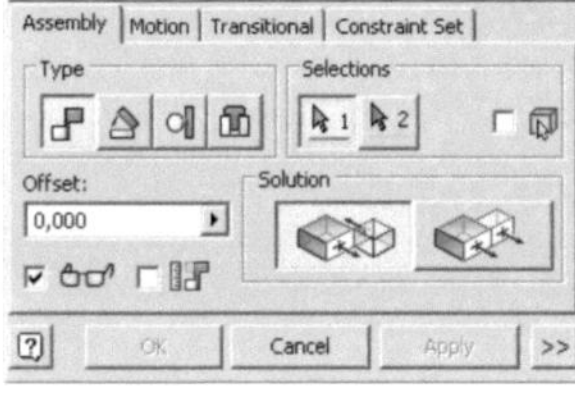

Now copy and paste the fitting key (**RMC** > **Copy**, **RMC** > **Paste**) and place it to the other side of the crankshaft.

7.2.3 Set depencies for components

After both fitting keys already have been mounted to the crankshaft the two pulleys need to get mounted.

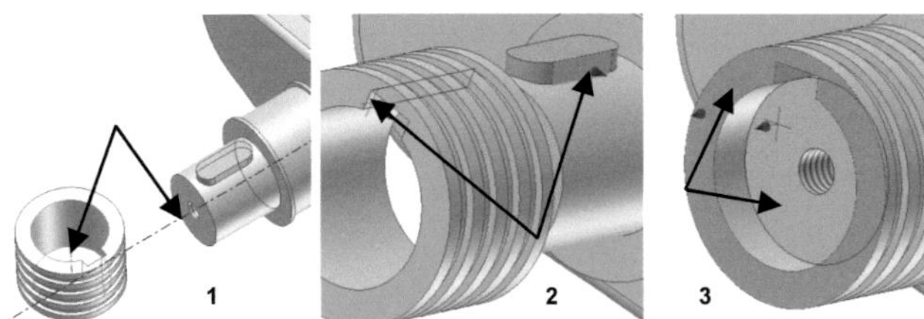

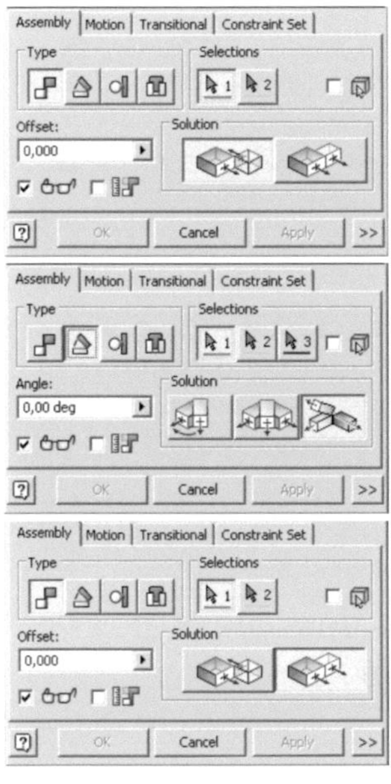

> ⬛ ***Constrain***
> Type: Mate
> Selections: Marked axes in picture 1
> Offset: 0
> Solution: Mate
> [Apply]
> Type: Angle
> Selections: Marked areas in picture 2
> Angle: 0
> Solution: 3
> [Apply]
> Type: Mate
> Selections: Marked areas in picture 3
> Offset: 0
> Solution: Flush
> [OK]

7.2.4   Locking washer

The locking washer, which secures the pulleys of the crankshaft against axial drifting, is created from the assembly group.

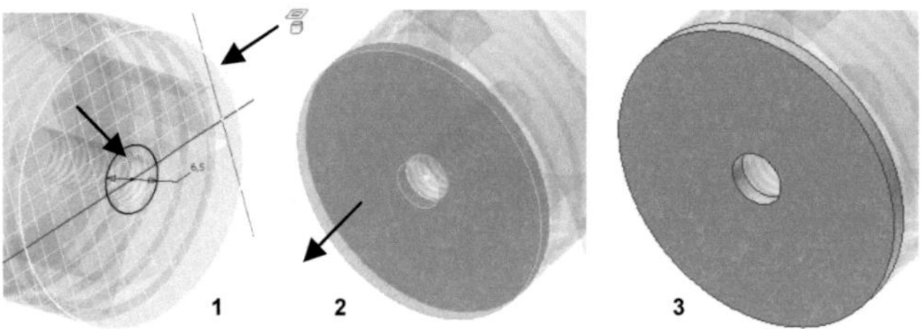

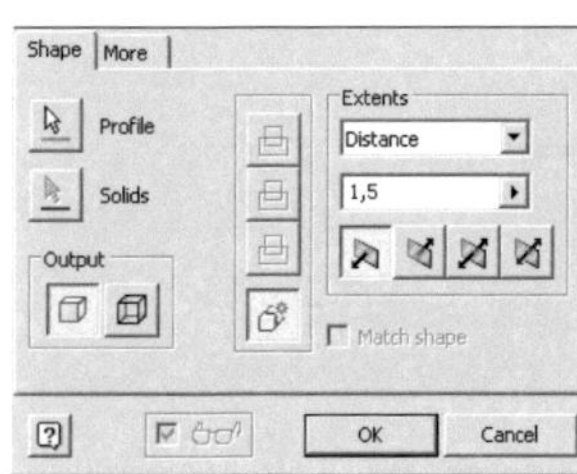

> □ *Create*
> New Component Name: locking-washer
> Template: Standard.ipt
> New File Location: Project save location
> Hook at: Constrain sketch plane to ...
> [OK]

> *Project Geometry* (marked outside edge, picture 1)
> Additionally draw a circle with D= 6.5 (concentric to projected circle)
> ✔ *Finish Sketch*

> *Extrude*
> Profile: Area between both circles (picture 2)
> Operation: Join
> Extents: Distance 1.5
> Direction: Shown (picture 2)
> [OK]

Tip: To use the just locking washer also on the other side we have the option to copy (***RMC – Copy***) and paste (***RMC – Paste***) it or re-import it into the assembly group with the command ***Place Component***. For this it is necessary to save it beforehand. This is done with the commands:

> 🖫 ***Safe** > **Save All***

Also save the newly created components (first saving).

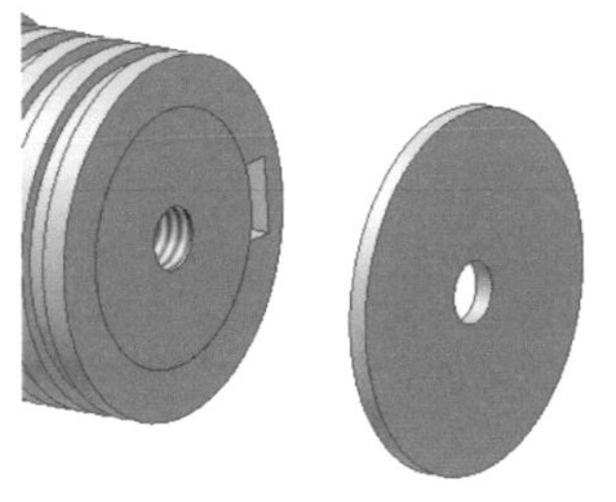

> 💾 *Save* > *Save all*

> 📦 *Place Component*
> Part: locking-washer.ipt

The locking washer is now to be mounted to the other side of the shaft (identical to the position of the first disk).

7.2.5　Hexagon screw from the content center

To tightly attach the locking washer with the crankshaft we also need two hexagon screws M6 x 14 from the content center.

Place the two screws fitting into the already existing threaded holes of the crankshaft.

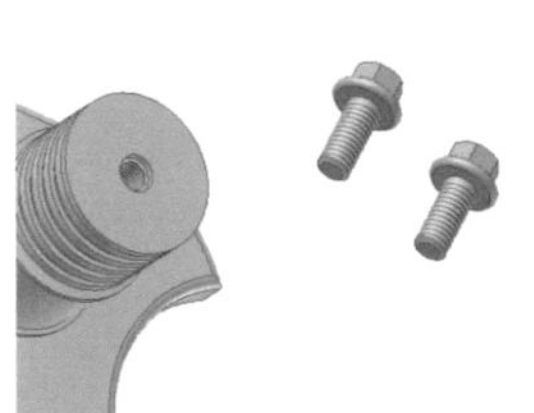

> 🖨 *Place from Content Center*
> Connecting Elements – Screws – Hexagon Screw with Flange – Hexagon Screw with Flange (metric)
> Thread Size: M6
> Nominal Length: 14
> ‹ OK ›

Note: If you are using *inch* units, please transform the metric ones to inch system.

Place the screw x 2 next to crankshaft.

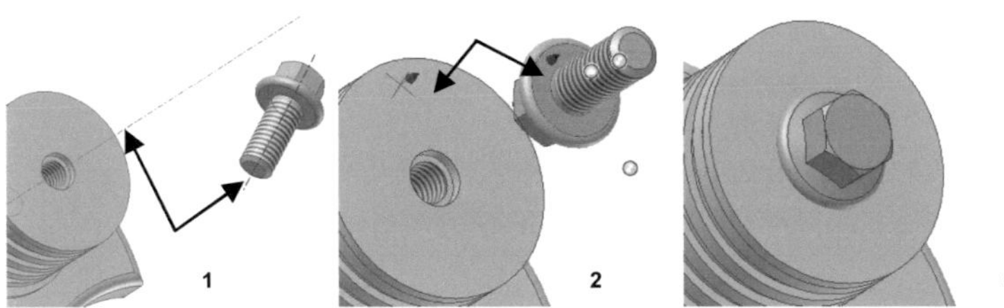

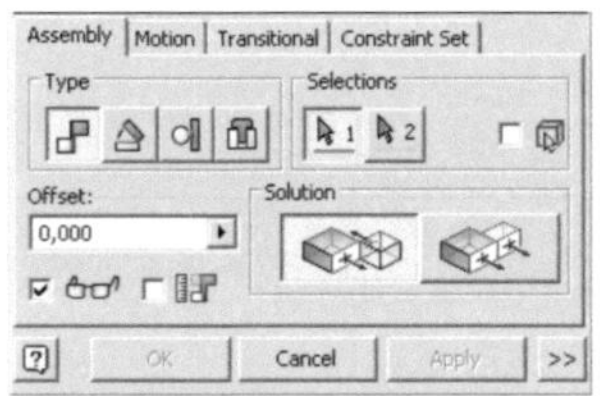

> ⊡ **Constrain**
> Type: Mate
> Selections: Marked axes in picture 1
> Offset: 0
> Solution: Mate
>   Apply
> Type: Mate
> Selections: Marked areas in picture 2
> Offset: 0
> Solution: Mate
>   OK

Then mount the second screw identical on the other side of the shaft.

7.2.6   Allocate colors

Then the components get **colors** allocated.

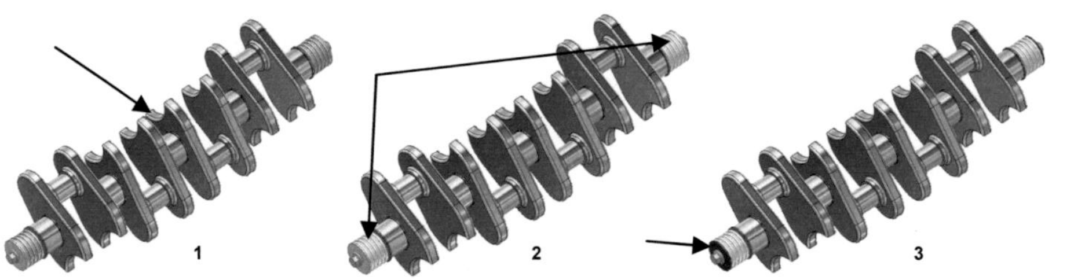

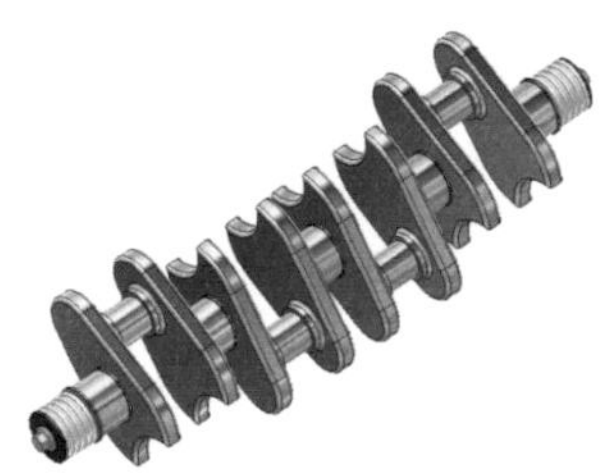

> Mark the **crankshaft** in the model tree
> **Color Override**
> Color: e.g. chrome (blue)
>
> Mark both **crankshaft pulleys** in the model tree
> **Color Override**
> Color: e.g. yellow
> Mark both **locking washers** in the model tree
> **Color Override**
> Color: e.g. red

assembly-crankshaft.iam now is complete. 💾 **Save** and ✖ **Close** the file.

7.3 Subassembly CAMSHAFT

Our next assembly group consists of 5 components (camshaft, camshaft-pulley, locking washer, fitting keys and hexagon screw).

Create a ▣ **New Assembly** and 🖫 **Save** it as **assembly-camshaft.iam** in your project folder.

7.3.1 Inserting the components

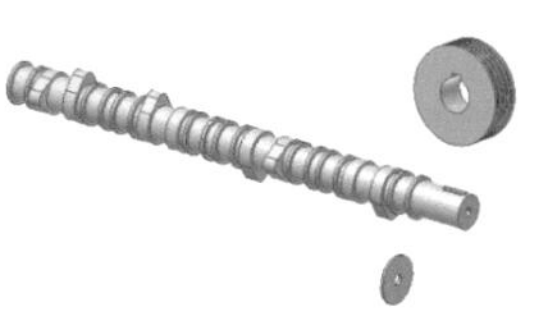

- ➢ 🖳 **Place component**
- ➢ 1x part: camshaft.ipt
- ➢ 1x part: camshaft-pulley.ipt
- ➢ 1x part: locking-washer. ipt

7.3.2 Locking washer from the content center

We need a fitting key DIN 6885 A 6 x 6 x 16, which will be used as connecting piece between shaft and pulley.

Insert these from the 🖨 **Content Center** and assign the shown constraints. Make sure to mount the fittings keys on both sides.

- ➢ 🖨 **Place from Content Center**
- ➢ Wave Parts – Fitting Keys – Fitting Key Rectangular – Rounded – DIN 6885 A
- ➢ Wave Diameter: 17-22
- ➢ Width x Height: 6 x 6
- ➢ Nominal Length: 16
- ➢ [OK]

Note: If you are using **inch** units, please transform the metric ones to inch system.

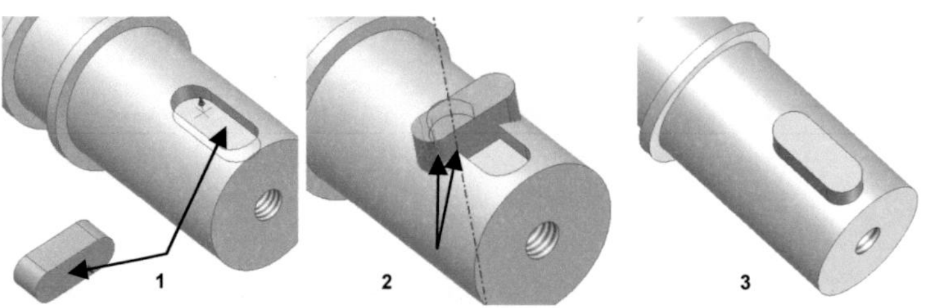

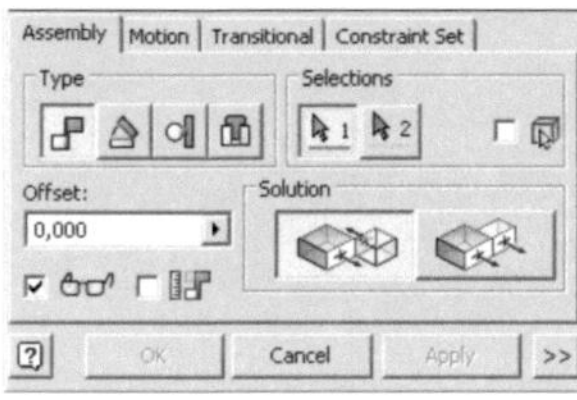

> 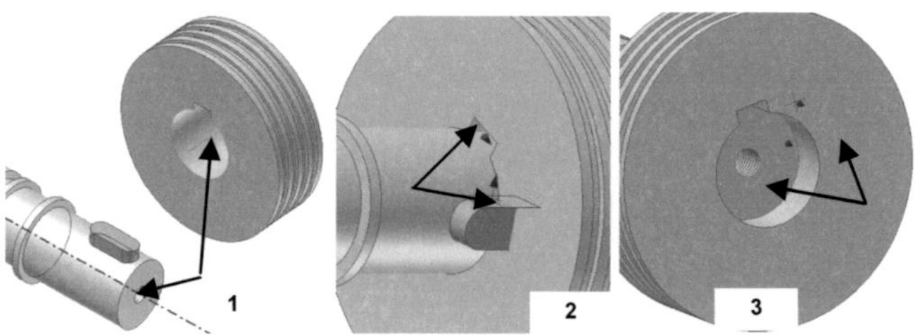 **Constrain**
> Type: Mate
> Selections: Marked areas in picture 1
> Offset: 0
> Solution: Mate
> Apply
> Type: Mate
> Selections: Marked axes in picture 2
> Offset: 0
> Solution: Mate
> Apply
> Type: Mate
> Selections: Axes on the other fitting key side
> Offset: 0
> Solution: Mate
> OK

7.3.3 Assign constraints

Mount the pulley of the camshaft axial to the shown wave end and place a **Constraint** between fitting key and fitting key notch on the pulley.

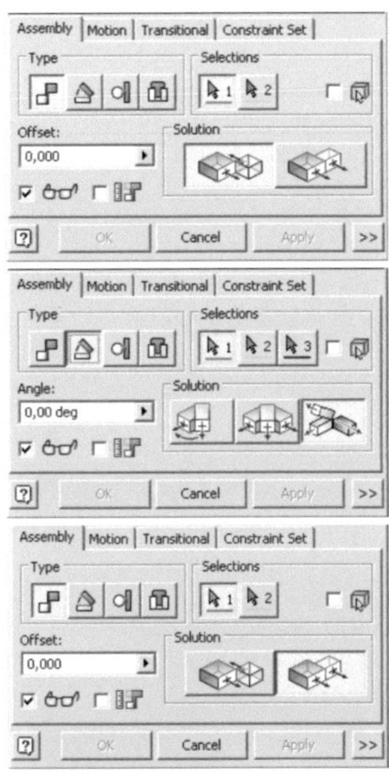

- ➢ ⌐ *Constrain*
- ➢ Type: Mate
- ➢ Selections: Marked axes in picture 1
- ➢ Offset: 0
- ➢ Solution: Mate
- ➢ [Apply]
- ➢ Type: Angle
- ➢ Selections: Marked areas in picture 2
- ➢ Angle: 0
- ➢ Solution: 1
- ➢ [Apply]
- ➢ Type: Mate
- ➢ Selections: Marked areas in picture 3
- ➢ Offset: 0
- ➢ Solution: Flush
- ➢ [OK]

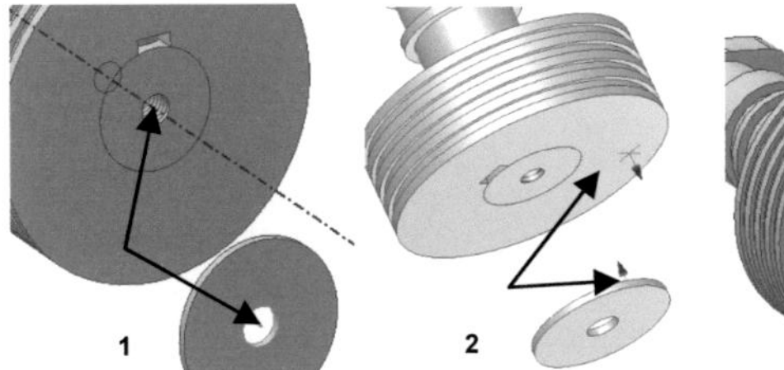

1   2   3

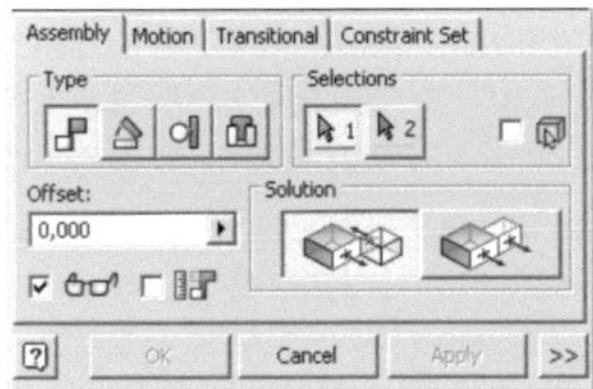

- ➢ ⌐ *Constrain*
- ➢ Type: Mate
- ➢ Selections: Marked axes in picture 1
- ➢ Offset: 0
- ➢ Solution: Mate
- ➢ [Apply]
- ➢ Type: Mate
- ➢ Selections: Marked areas in picture 2
- ➢ Offset: 0
- ➢ Solution: Mate
- ➢ [OK]

7.3.4 Hexagon screw from the content center

To secure the components we create a hexagon screw M6 x 14 from the 🖶 **Content Center** and place it.

> ➢ 🖶 *Place from Content Center*
> ➢ Connective Elements – Screws – Hexagon Screws with Flange – Hexagon Screw with Flange (metric)
> ➢ Thread Size: M6
> ➢ Nominal Length: 14

Note: If you are using *inch* units, please transform the metric ones to inch system.

Place the created screw 1 x into the drawing window.

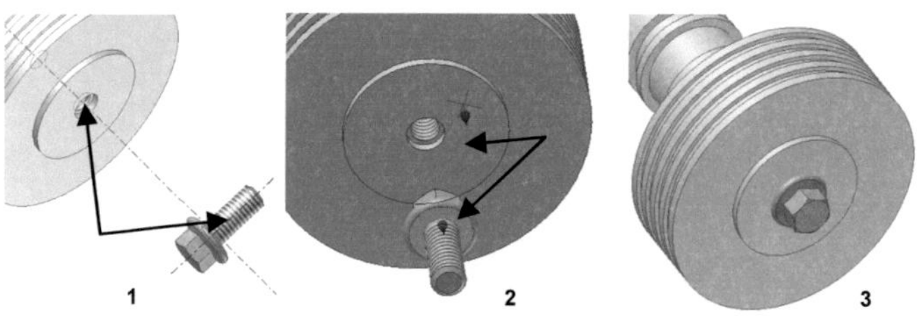

1   2   3

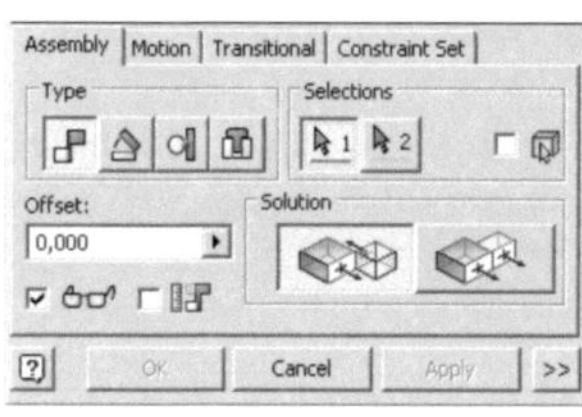

> ➢ ⬚ *Constrain*
> ➢ Type: Mate
> ➢ Selections: Marked axes in picture 1
> ➢ Offset: 0
> ➢ Solution: Mate
> ➢ [Apply]
> ➢ Type: Mate
> ➢ Selections: Marked areas in picture 2
> ➢ Offset: 0
> ➢ Solution: Mate
> ➢ [OK]

7.3.5   Assign colors

Assign colors to the components.

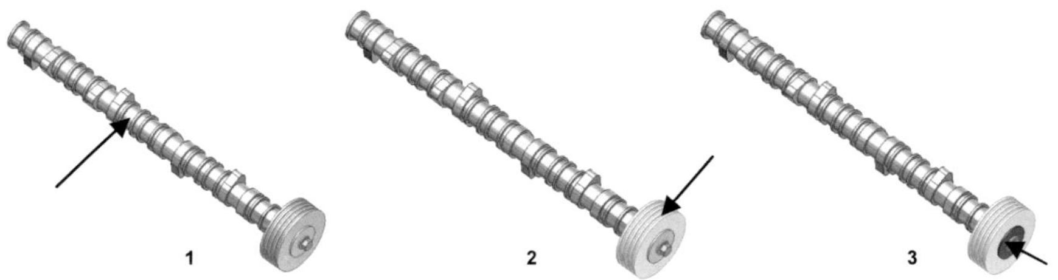

> ➢ Mark the **camshaft** in the model tree
> ➢ _Farbe_ **Color Override**
> ➢ Color: e.g. chrome (blue)

> ➢ Mark **camshaft-pulley** in the model tree
> ➢ _Farbe_ **Color Override**
> ➢ Color: e.g. yellow

> ➢ Mark **locking washer** in the model tree
> ➢ _Farbe_ **Color Override**
> ➢ Color: e.g. red

assembly-camshaft.iam now is complete. 🖫 **Save** and ✖ **Close** the file.

7.4   Subassembly CYLINDER BLOCK

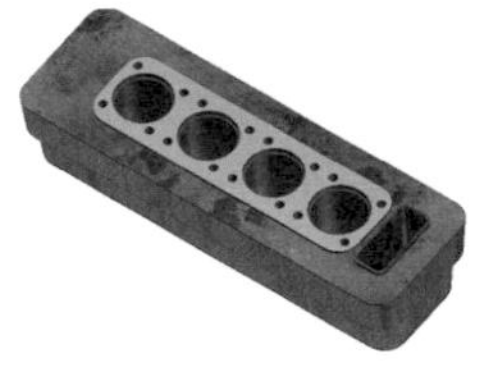

The assembly engine block contains 5 components; the engine block and the 4 sleeves.

Create a 🗋 **New Assembly** and 🖫 **Save** it as **assembly-cylinder-block.iam** in your project folder.

7.4.1   Inserting the components

🖳 **Insert** the two components **cylinder block** and **sleeves** and then place the marked constraints.

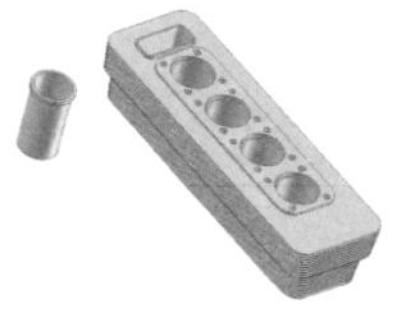

> 📇 ***Place Component***
> ➢ 1x part: cylinder-block.ipt
> ➢ 1x part: sleeve.ipt

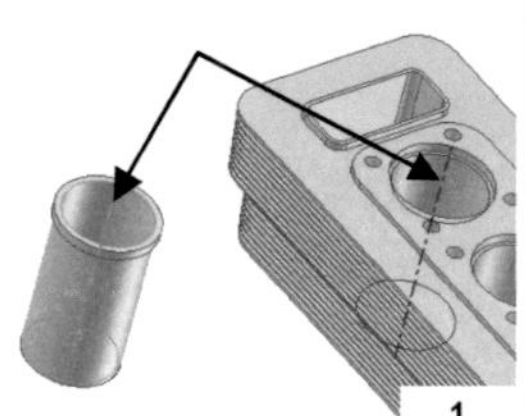
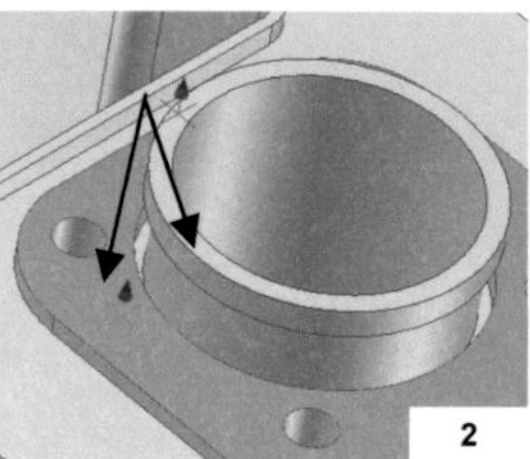
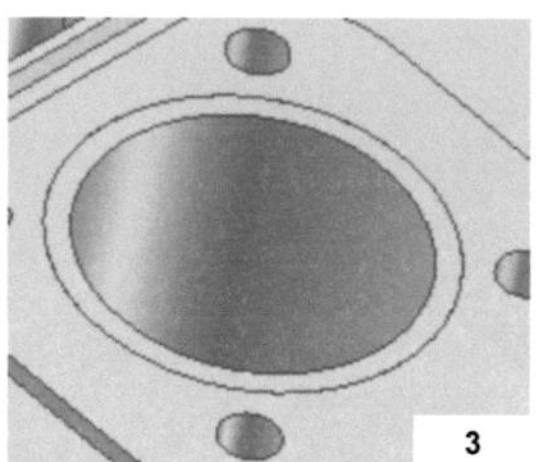

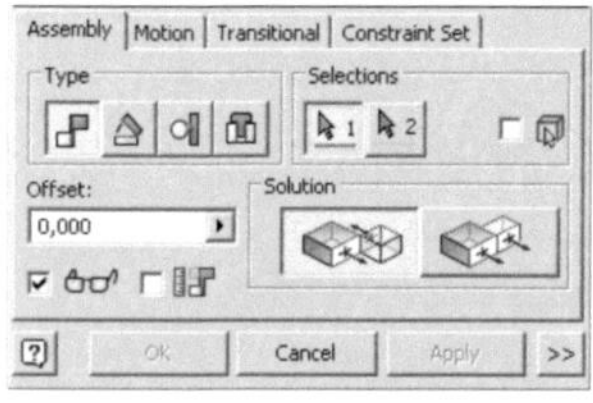
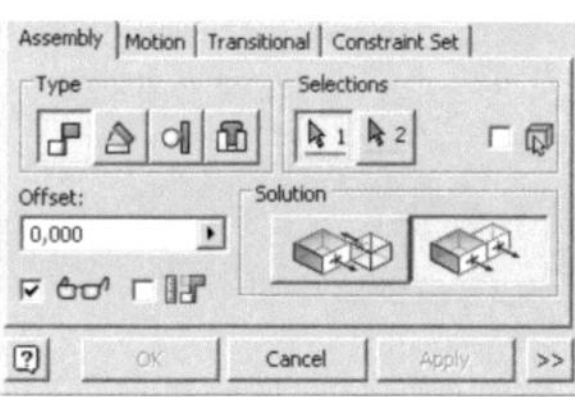

> ⬚ ***Constrain***
> ➢ Type: Mate
> ➢ Selections: Marked axes (picture 1)
> ➢ Offset: 0
> ➢ Solution: Mate
> ➢ [Apply]
> ➢ Type: Mate
> ➢ Selections: Marked areas (picture 2)
> ➢ Offset: 0
> ➢ Solution: Flush
> ➢ [OK]

7.4.2 Organizing the sleeves

The sleeve of the first cylinder is positioned. With the command ⬚ ***Pattern Component*** we create and positions the remaining 3 sleeves.

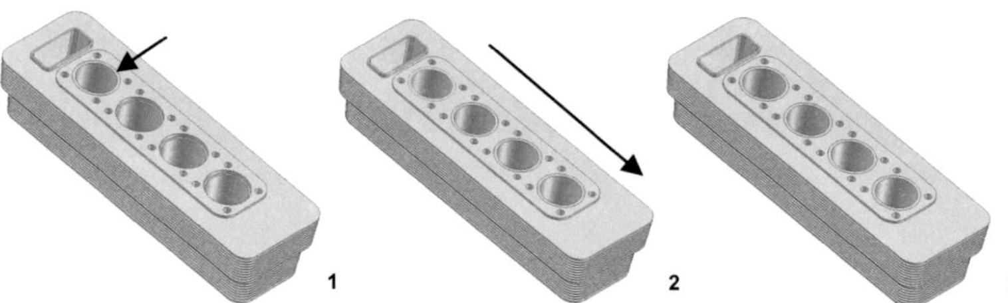

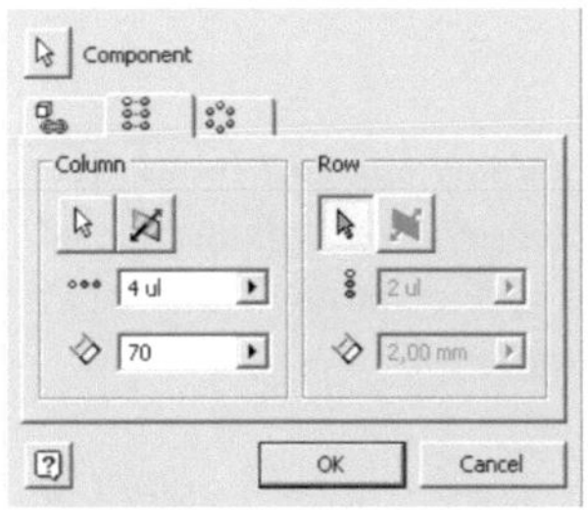

- ➢ ⸬ **Pattern Component**
- ➢ Elements: Sleeve
- ➢ Direction: Z-axis
- ➢ Quantity: 4
- ➢ Distance: 70
- ➢ [OK]

7.4.3 Assign colors

Assign different colors to the components colors.

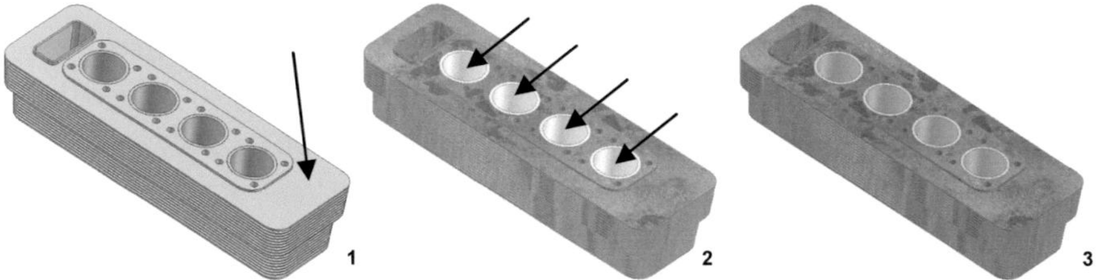

- ➢ Mark the **cylinder block** in the model tree
- ➢ _Farbe_ **Color Override**
- ➢ Color: e.g. steel metal (rusted)

- ➢ Mark 4 **sleeves** in the model tree
- ➢ _Farbe_ **Color Override**
- ➢ Color: e.g. steel metal (honed)

assembly-cylinder-head.iam now is complete 💾 **Save** and ✕ **Close** the file.

Tip: Should it be necessary to assign different color to separate areas within a part you can assign them directly in editing the part. Open, e.g. a part, choose an area and click:

➢ **RMC** > **Properties**

To assign the desired color (area color style). Please consider: When you assign a new color to the entire part in the assembly group, the color assigned to the part are no longer shown. Upper sting lower, applies here.

7.5   Subassembly CYLINDER HEAD

Our last subassembly is the assembly cylinder head. This subassembly consists of the components cylinder head, spark plug, camshaft bracket, Allen screw and shaft sealing ring from the content center.

Create a ▣ **New Assembly** and save it as ***assembly-cylinder-head.iam*** in your project folder.

7.5.1   Inserting the components

Insert the shown components and then assign all needed constraints.

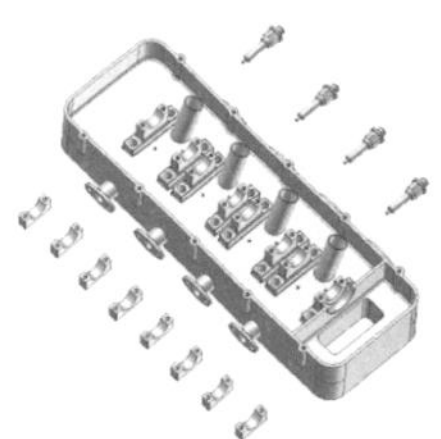

> ➢ 🖳 **Place Component**
> ➢ 1x part: cylinder-block.ipt
> ➢ 8x part: valve.ipt
> ➢ 8x part: camshaft-holder.ipt
> ➢ 4x part: sparkplug.ipt

Place the components into the drawing window.

7.5.2   Assign constraints

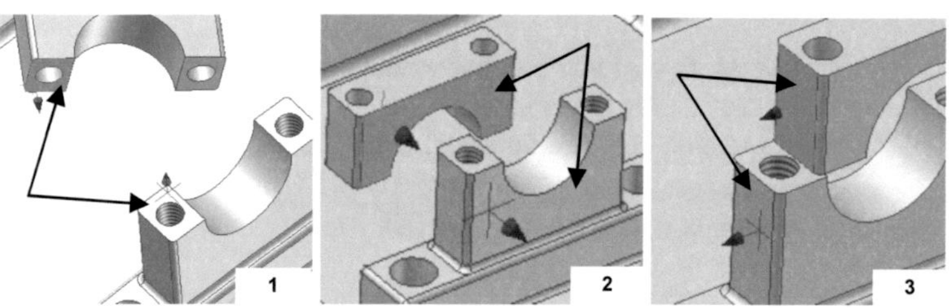

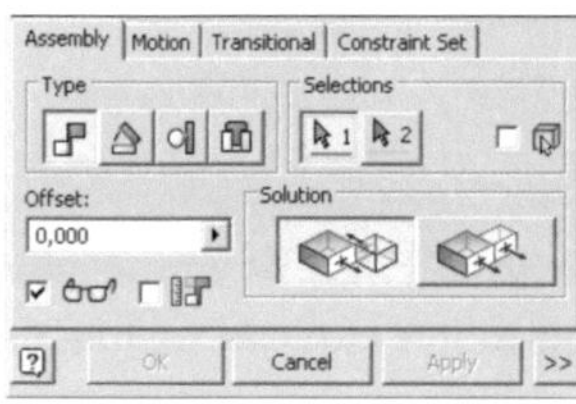

> ➢ ⬚ **Constrain**
> ➢ Type: Mate
> ➢ Selections: Marked areas (picture 1)
> ➢ Offset: 0
> ➢ Solution: Mate
> ➢ [Apply]
> ➢ Type: Mate
> ➢ Selections: Marked areas (picture 2)
> ➢ Offset: 0

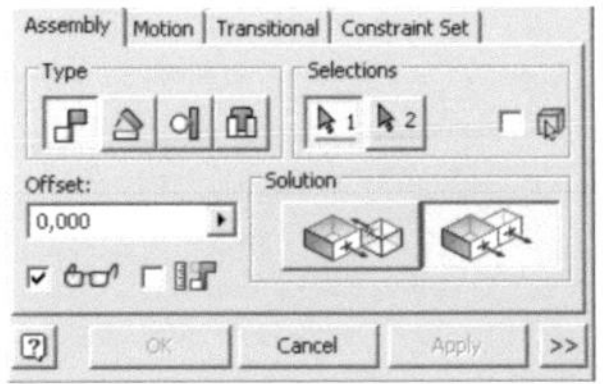

➢ Solution: Flush
➢ [Apply]
➢ Type: Mate
➢ Selections: Marked areas (picture 3)
➢ Offset: 0
➢ Solution: Flush
➢ [OK]

Repeat command for the remaining 7 camshaft holders.

Tip: It is also possible to fix the 2nd camshaft holder the same way and then create the remaining 6 camshaft holders by using the command 🔳 **Pattern Component** (quantity: 4, distance: 70).

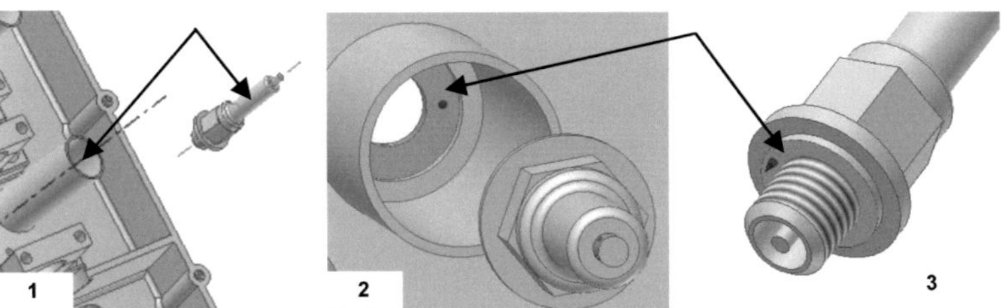

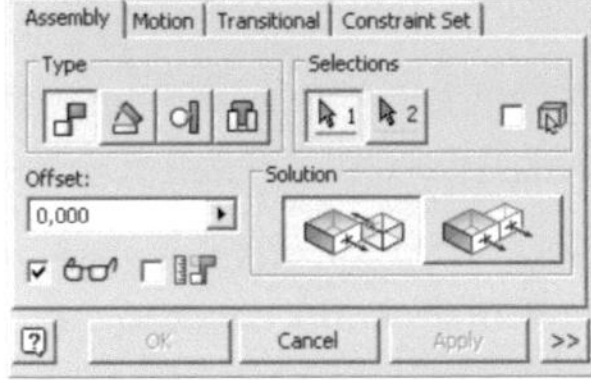

➢ 🔳 **Constrain**
➢ Type: Mate
➢ Selections: Marked axes (picture 1)
➢ Offset: 0
➢ Solution: Mate
➢ [Apply]
➢ Type: Mate
➢ Selections: Marked areas (picture 2+3)
➢ Offset: 0
➢ Solution: Mate
➢ [OK]

Repeat the commands for the remaining 3 spark plugs.

Tip: Here it is also possible to create the remaining 3 spark plugs with the command 🔳 **Pattern Component** (quantity: 4, distance: 70).

7.5.3 Assign colors

Now we will assign colors to the available components.

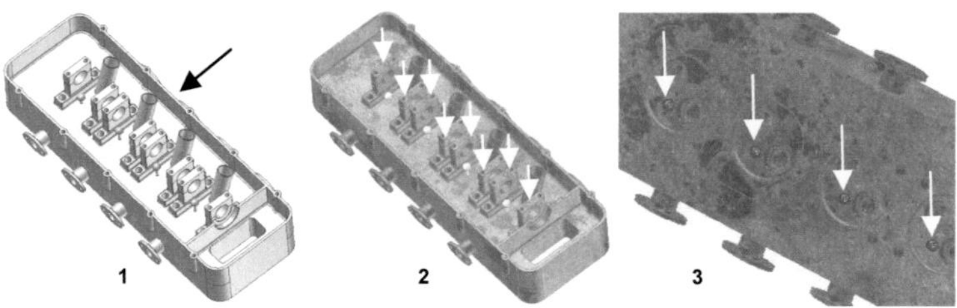

> ➤ Mark **cylinder head** and the 8 **camshaft holders** in the model tree
> ➤ *Farbe* **Color Override**
> ➤ Color: e.g. steel metal (rusted)
>
> ➤ Mark the 4 **spark plugs** in the model tree]
> ➤ *Farbe* **Color Override**
> ➤ Color: e.g. red

Tip: When you created the components **camshaft holder** and **spark plug** by using the command ⠿ **Pattern Component** you will find the components under the respective component orders.

7.5.4 Allen screws from the content center

Then the connection screws between cylinder head and camshaft holder are created from the 🖶 **Content Center**. Here we use a cylinder head screw M6 x 25.

When you defined the first screw you can repeat the commands for the remaining 15 screws or work with the command ⠿ **Pattern Component**.

DIN EN ISO 1207

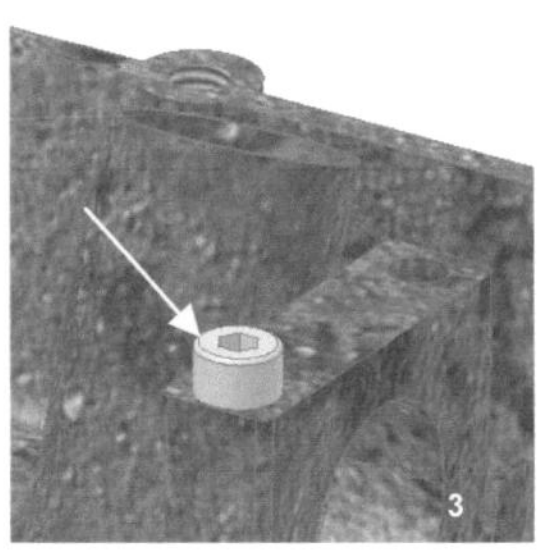

> ➤ 🖶 *Place from Content Center*
> ➤ Connection Elements – Screws – Cylinder Head – DIN EN ISO 4762
> ➤ Thread Size: M6
> ➤ Nominal Length: 25
> ➤ [OK]

Note: If you are using *inch* units, please transform the metric ones to inch system.

Place the created screw 1x next to the cylinder head and then place as shown.

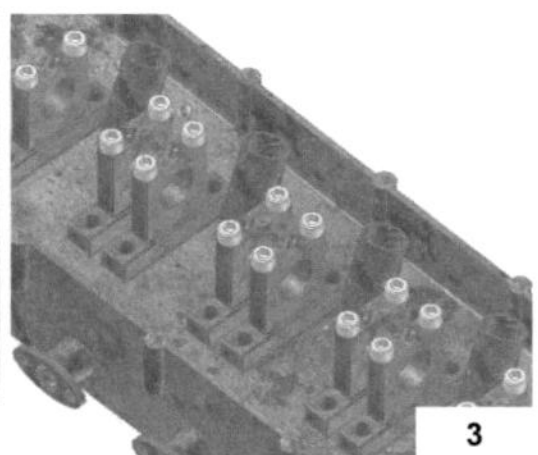

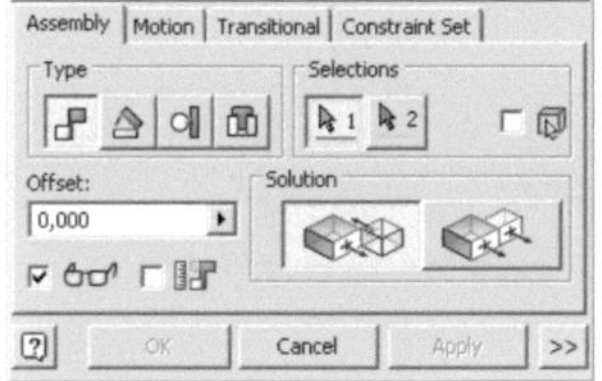

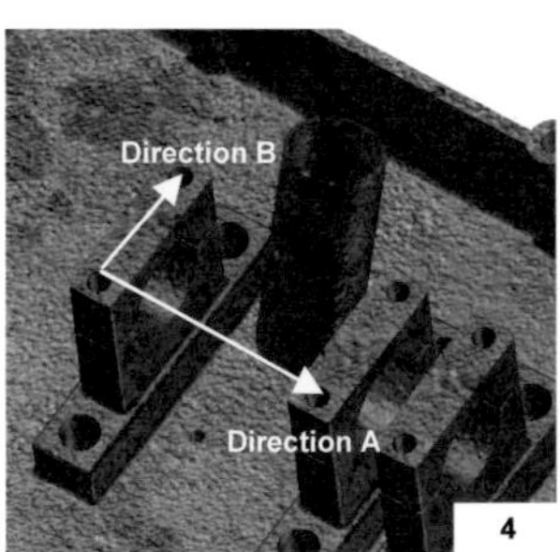

> ➤ ⬚ *Constrain*
> ➤ Type: Mate
> ➤ Selections: Marked axes (picture 1)
> ➤ Offset: 0
> ➤ Solution: Mate
> ➤ [Apply]
> ➤ Type: Mate
> ➤ Selections: Marked areas (picture 2)
> ➤ Offset: 0
> ➤ Solution: Mate
> ➤ [OK]

Now we place the remaining 15 screws. Use the command ⬚ *Pattern Component* in two steps.

For the directions A and B use one of the available edges.

> ⊞ **Pattern Component** (Step 1)
> Elements: First screw
> Direction 1: Direction A (picture 4)
> Quantity 1: 2
> Distance 1: 50
> Direction 2: Direction B (picture 4)
> Quantity 2: 2
> Distance 2: 27.5
> [OK]

> ⊞ **Pattern Component** (Step 2)
> Elements: Last pattern from step 1
> Direction 1: Direction A (picture 4)
> Quantity 1: 4
> Distance 1: 70
> [OK]

7.5.5 Shaft sealing ring from the content center

The camshaft space has to get sealed towards the belt space. For this we use a sealing ring from the 🖶 **Content Center**. Then place the sealing ring as shown.

ANSI/B93.98M

ANSI/B93.98M

ANSI/B93.98M (4)

DIN 3760 A

> 🖶 **Place from Content Center**
> Shaft Parts – Seals – Sealing Rings – DIN 3760 A
> Tab: Table View
> Line Status: 41 (20 x 35 x 7)
> [OK]

Note: If you are using *inch* units, please transform the metric ones to inch system.

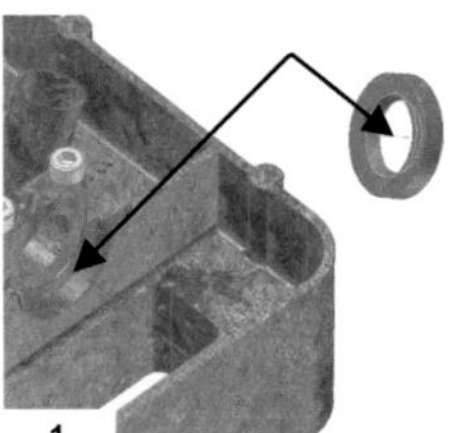

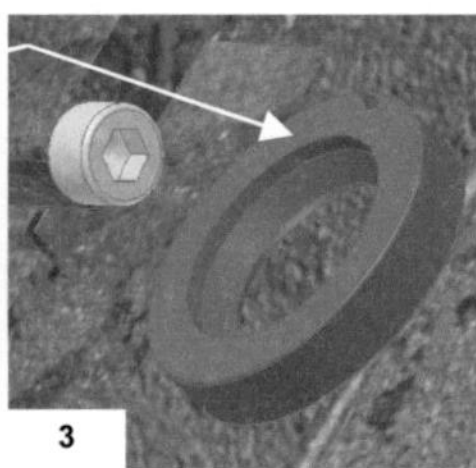

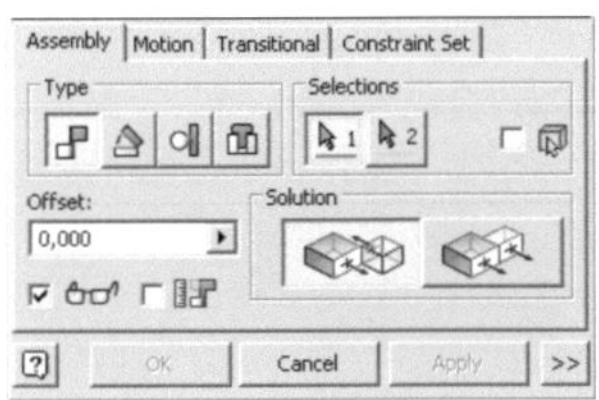

> ⌐ *Constrain*
> Type: Mate
> Selections: Marked axes (picture 1)
> Offset: 0
> Solution: Mate
> | Apply |
> Type: Mate
> Selections: Marked areas (picture 2+3)
> Offset: 0
> Solution: Mate
> | OK |

assembly-cylinder-head.iam now is complete. ⊟ *Save* and ✕ *Close* the file.

7.6 Main assembly group *FOUR-STROKE-ENGINE*

Our subassemblies have successfully been created, now we can start on the main assembly.

The main assembly consists of the components assembly crankshaft, assembly cylinder head, assembly camshaft, assembly piston, engine block, valve cover, Timing belt, valve, cylinder head screw and composition.

7.6.1 Inserting crankshaft & piston

Create a ▣ *New Assembly* and ⊟ *Save* it in the project folder as ***assembly-four-stroke-engine.iam***. Then insert following components:

> ▤ *Place Component*
> 1x assembly: assembly-crankshaft.iam
> 4x assembly: assembly-piston.iam

Place the components into the drawing window.

7.6.2 Connect crankshaft with pistons

⌐ *Place* the first piston-assembly onto the crankshaft. Make sure to leave distance of 1 on the faces between piston rod and crankshaft (picture 4).

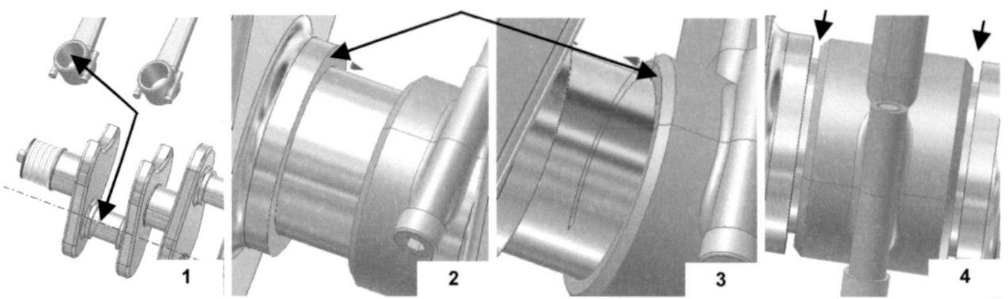

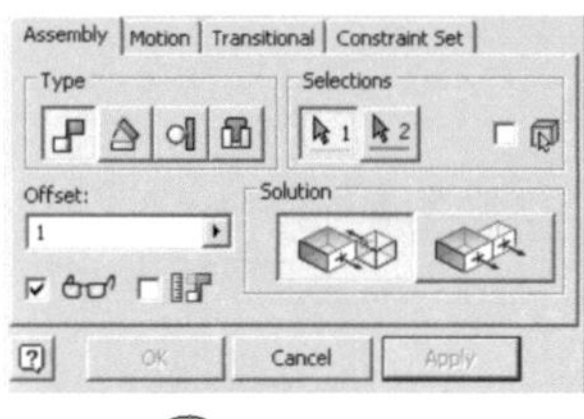

> ⬦ **Constrain**
> Type: Mate
> Selections: Marked axes (picture 1)
> Offset: 0
> Solution: Mate
> Apply
> Type: Mate
> Selections: Marked areas (picture 2+3)
> Offset: 1
> Solution: Mate
> OK

Repeat constraints for the remaining 3 pistons.

Tip: When subassemblies are inserted into assemblies they are declared **non-flexible** by default (they are treated as a solid part). Since we however, need a movable piston for our assembly we have to define the 4 pistons to be **flexible**.

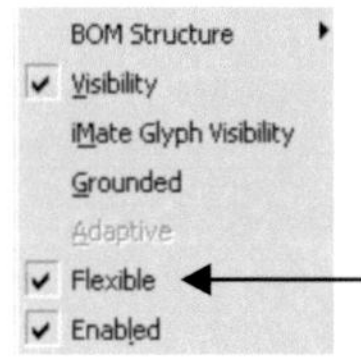

> Click with **RMC** on the 4 pistons
> Mark pistons as **Flexible**

7.6.3   Insert engine block & provide with assemblies

Insert the component **engine-block.ipt** into the assembly and place the crankshaft-piston-group into it. The axial location of the crankshaft is determined by the correct positioning of the crankshaft in the engine block. The pistons are axially aligned to the holes for the liners.

The existing air gap between the two components will later be closed by the sleeves which also reach a bit into the engine block. After the components are aligned, we will make use of the properties of fixation of parts. Fixate the engine block (**RMC** > **Grounded**) and rotate the crankshaft a little.

The pistons should move linear to it. If we skip the step of fixation we would move all parts in the space instead of having a rotational movement of the crankshaft.

> 🗋 **Place Component**
> 1x part: engine-block.ipt

> Mark the **engine block** in the model tree
> 🎨 **Color Override**
Color: e.g. steel metal (rusted)

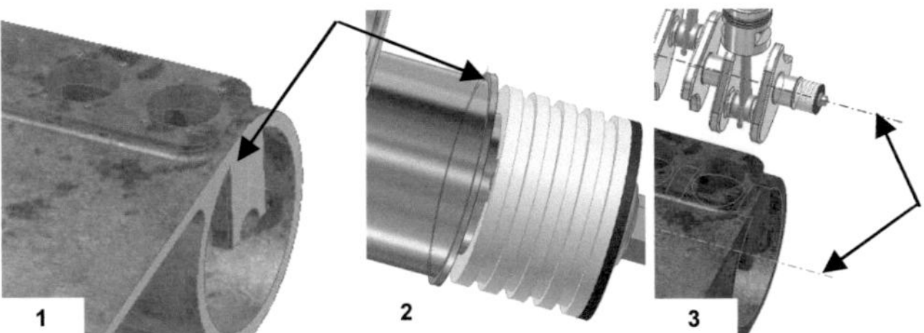

> ⬛ **Constrain**
> Type: Mate
> Selections: Marked areas (picture 1+2)
> Offset: 1
> Solution: Mate
> Apply
> Type: Mate
> Selections: Marked axes (picture 3)
> Offset: 0
> Solution: Mate
> OK

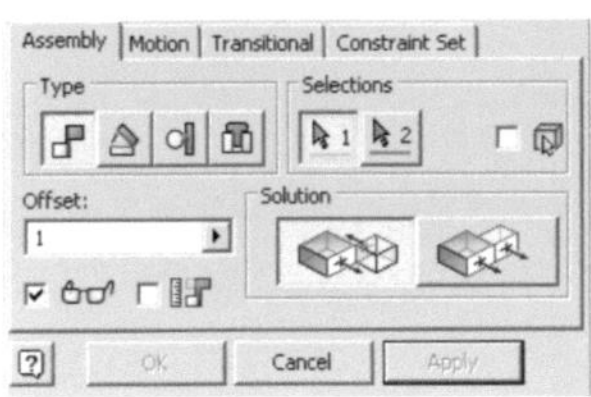

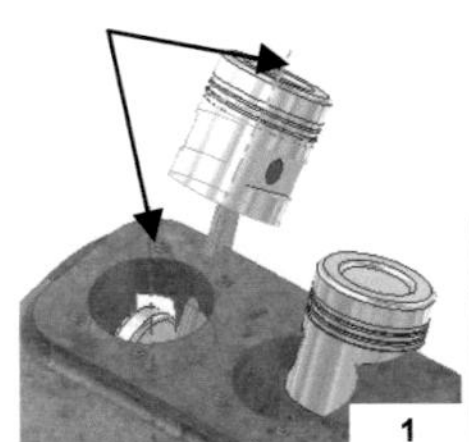

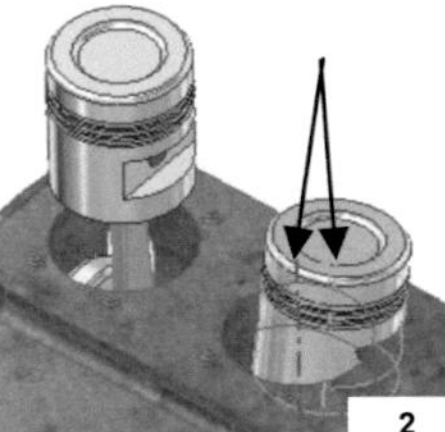

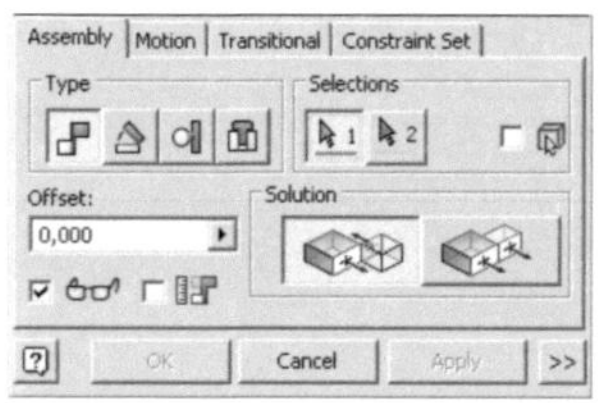

> ⬚ ***Constrain***
> ➢ Type: Mate
> ➢ Selections: Marked axes (picture 1)
> ➢ Offset: 0
> ➢ Solution: Mate
> ➢ ⬚ OK

Place all 4 pistons in an axial constraint to the holes in the motor case (picture 1+2)

Tip: It is useful to fixate the engine block (**RMC** > **Grounded**) to start a test run. Now slightly rotate the crankshaft at pressed **LMC,** the 4 pistons should move linear to it.

Insert the component ***crankshaft-holder.ipt***, place it as shown and assign its own color.

> ➢ 🖳 ***Place Component***
> ➢ 1x part: crankshaft-holder.ipt
>
> ➢ Mark ***crankshaft-holder*** in the model tree
> ➢ 🎨 ***Color Override***
> ➢ Color: e.g. steel metal (rusted)

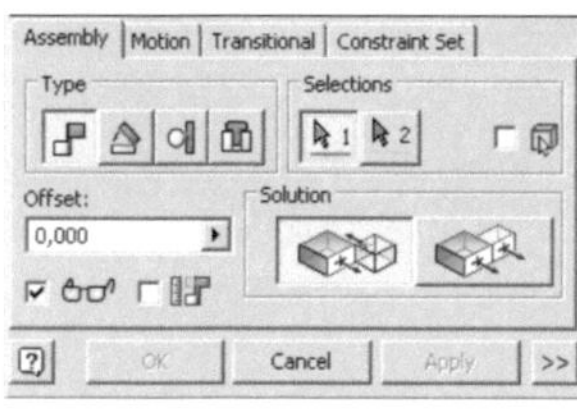

> ➢ ⬚ ***Constrain***
> ➢ Type: Mate
> ➢ Selections: Marked areas (picture 1)
> ➢ Offset: 0
> ➢ Solution: Mate
> ➢ Apply
> ➢ Type: Mate
> ➢ Selections: Marked axes (picture 2)
> ➢ Offset: 0
> ➢ Solution: Mate
> ➢ Apply
> ➢ Type: Mate

> ➢ Selections: Marked axes (picture 3)
> ➢ Offset: 0
> ➢ Solution: Mate
> ➢   OK

To mount the crankshaft holder to the engine block we use two screws DIN EN ISO 4762 from the 🖨 **Content Center**.

With the command ⊞ **Place more** both screws are set at the same time. Then we change the length of the screws.

DIN EN ISO 1207

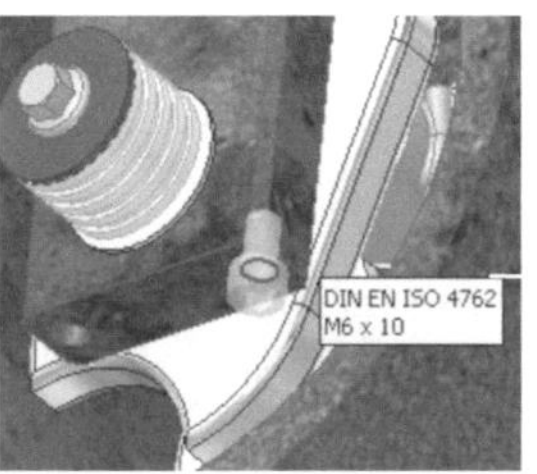

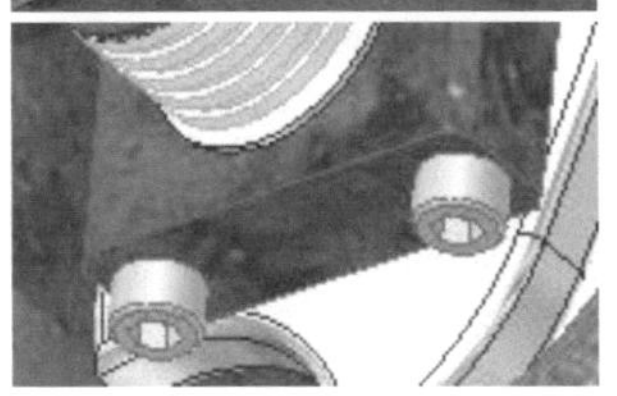

> ➢ 🖨 **Place from Content Center**
> ➢ Connection Elements – Screws – Cylinder Head – DIN EN ISO 4762
> ➢ Place **LMC** on marked hole edge (picture 3)
> ➢ The program automatically chooses the fitting diameter
>
> ➢ ⊞ **Place more**
> ➢ The program automatically finds the second threaded hole
> ➢ ✓

Note: If you are using **inch** units, please transform the metric ones to inch system.

> ➢ In the model tree click **RMC** on the just created screws > **Edit Size**
> ➢ New size: M6 x 35 (change both screws!)

Use the command ⊞ **Pattern Component**, to create the remaining 4 crankshaft holders including screws.

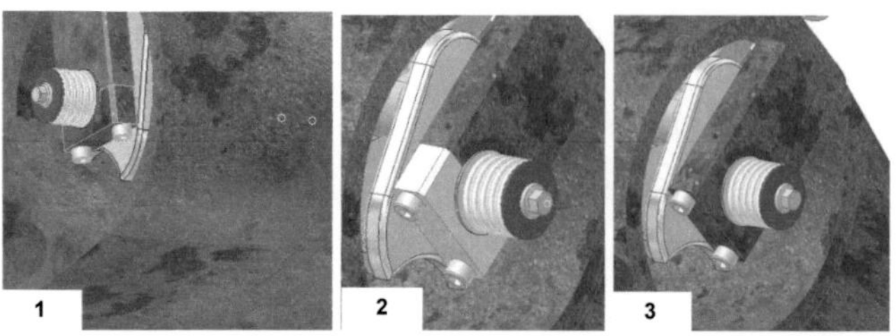

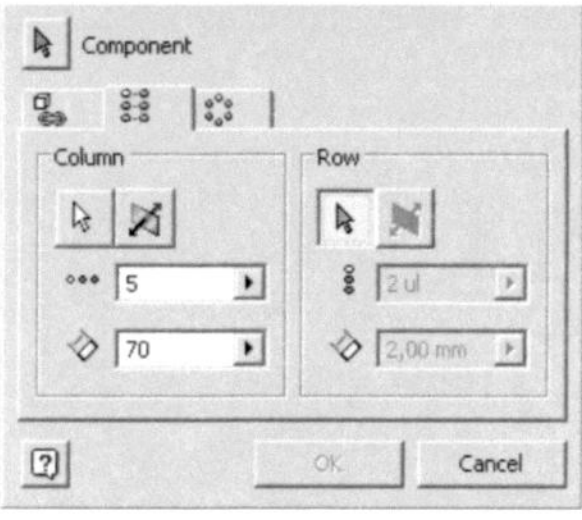

> ⊞ **Pattern Component**
> ➤ Components: Crankshaft holder, both screws
> ➤ Direction: Z-axis of the engine block
> ➤ Quantity: 5
> ➤ Distance: 70
> ➤ [OK]

7.6.4   Create part cylinderhead-composition

Before we can insert the assembly cylinder head we have to create a composition. The composition is located on the composition surface between engine block and cylinder head.

Create a new part on the marked area, project all edges of the composition surface and close the sketch.

Make sure that also the threaded holes for the cylinder head screws and the holes for the sleeves are marked (by clicking on the area all necessary lines are automatically marked).

Extrude the area and then leave the editing part of the seal.

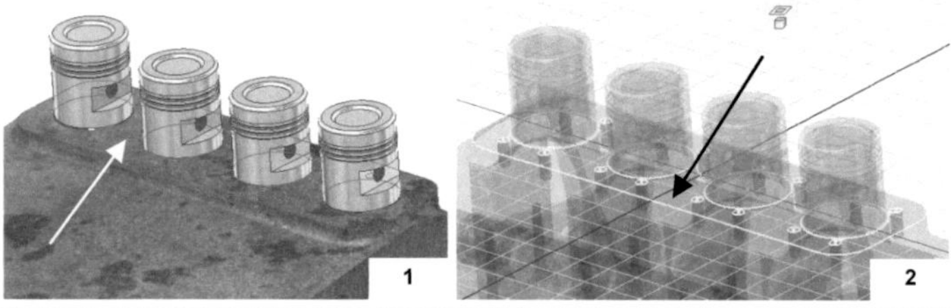

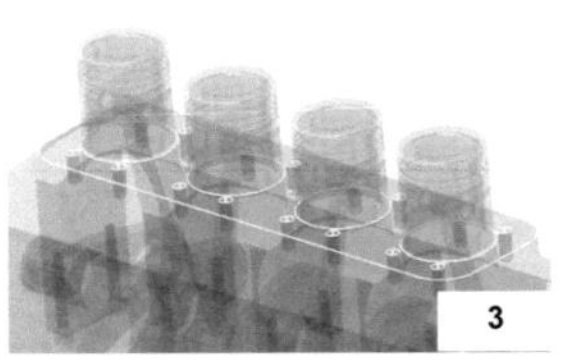
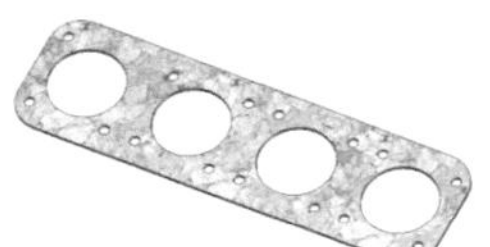

> ✐ **Create**
> Name: cylinderhead-composition.ipt
> Position: Marked area (picture 1)
> Template: Norm.ipt
> File saving location: project saving location
> Hook at: …. Assign constraint
> [OK]

> ✑ **Project Geometry** (all edges of the marked area in picture 2)
> ✔ **Finish Sketch**

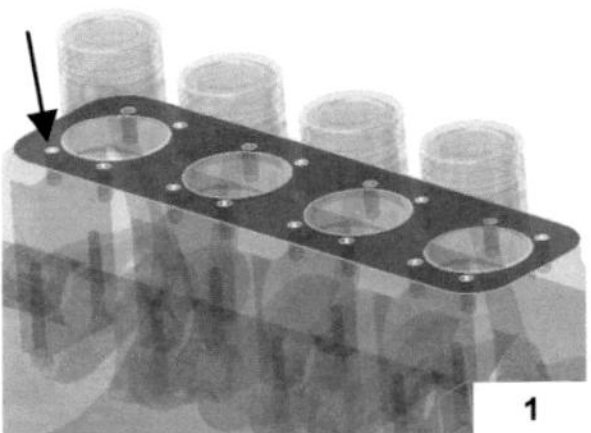
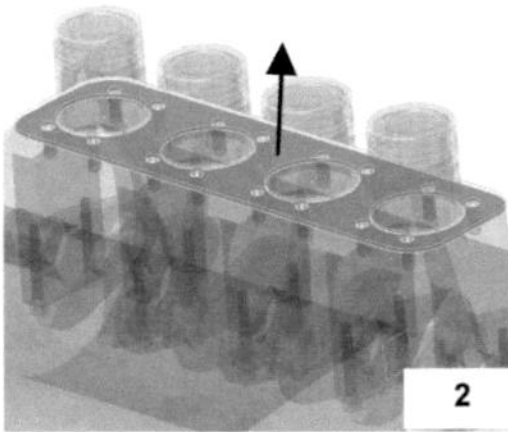
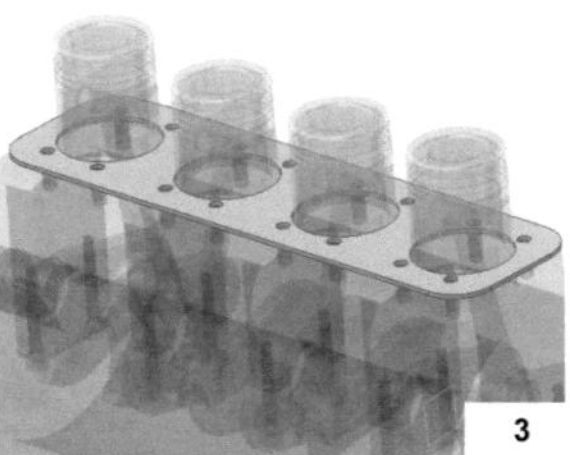

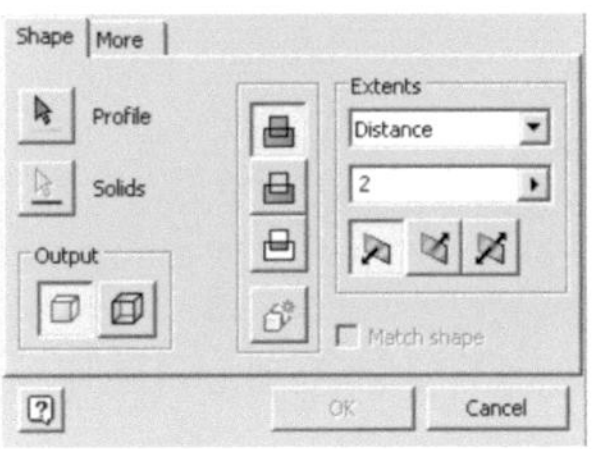

> ◻ **Extrude**
> Profile: Marked area (picture 1)
> Operation: Join
> Extents: Distance 2
> Direction: Shown (picture 2)
> [OK]

← **Back**

> Mark **cylinder head composition** in the model tree
> ▦ **Color Override**
> Color: e.g. cork

7.6.5   Inserting the cylinder head

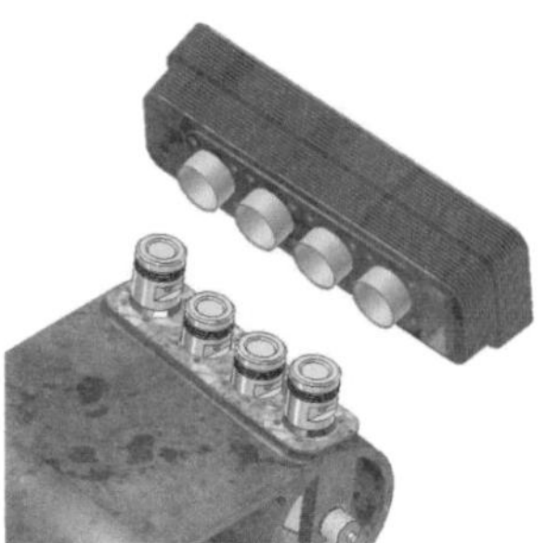

After we created the composition we now can insert the assembly cylinder head.

Place it as shown.

- ➤ 🖳 ***Place Component***
- ➤ 1x part: assembly-cylinder-head.iam

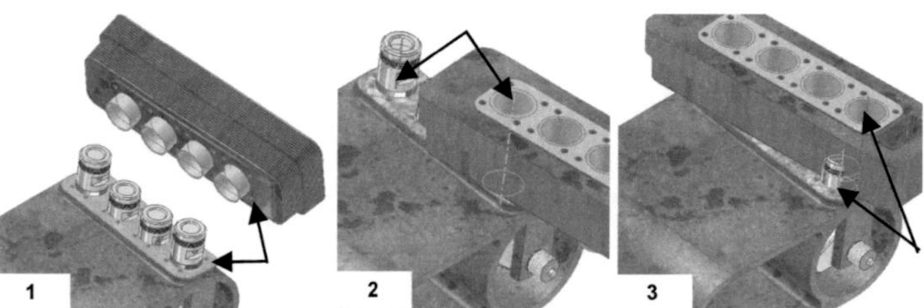

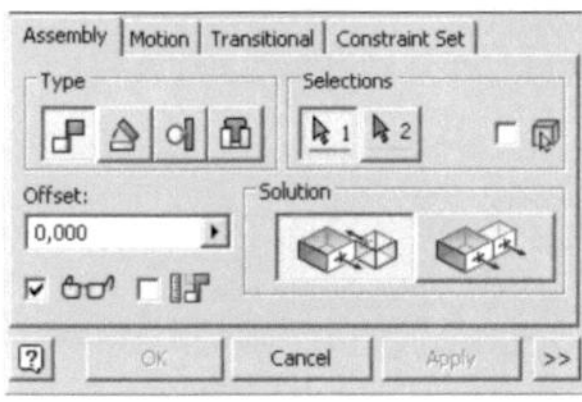

- ➤ ⬚ ***Constrain***
- ➤ Type: Mate
- ➤ Selections: Marked areas (picture 1)
- ➤ Offset: 0
- ➤ Solution: Mate
- ➤ Apply
- ➤ Type: Mate
- ➤ Selections: Marked axes (picture 2)
- ➤ Offset: 0
- ➤ Solution: Mate
- ➤ Apply
- ➤ Type: Mate
- ➤ Selections: Marked axes (picture 3)
- ➤ Offset: 0
- ➤ Solution: Mate
- ➤ OK

7.6.6   Create part cylinder head composition

We need a further composition between engine block and cylinder head.

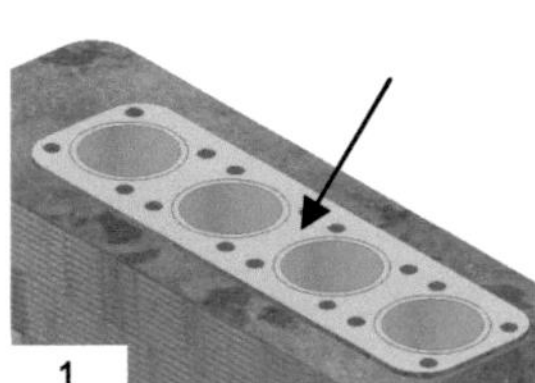

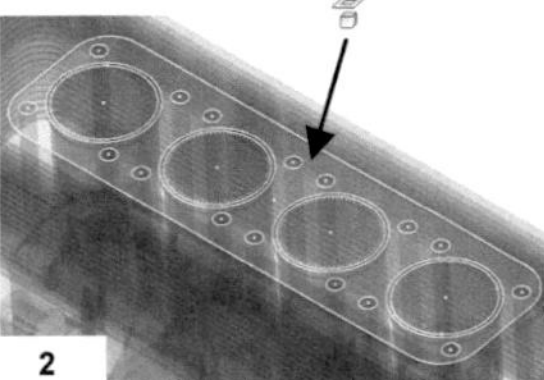

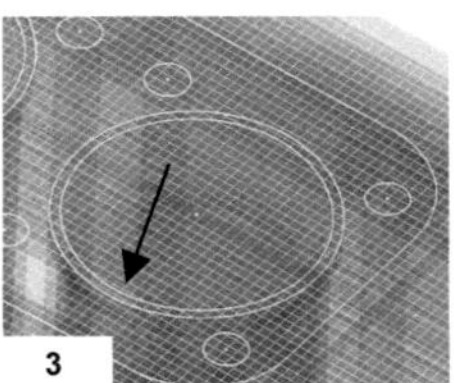

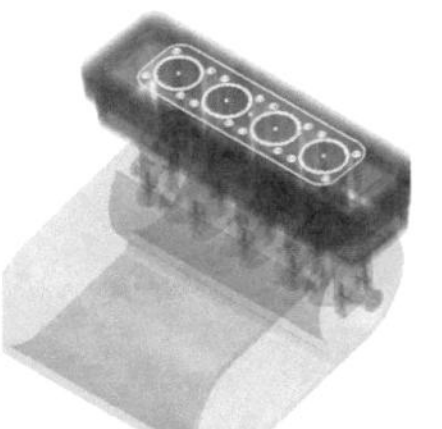

> ⬡ **Create**
> Name: cylinder-head-composition.ipt
> Position: Marked area (picture 1)

> ⬡ **Project Geometry**
> Project all edges of the face marked in picture 2 (don´t project the 4 circles D= 47, or mark them as ⅄ **Construction Line**
> ✔ **Finish Sketch**

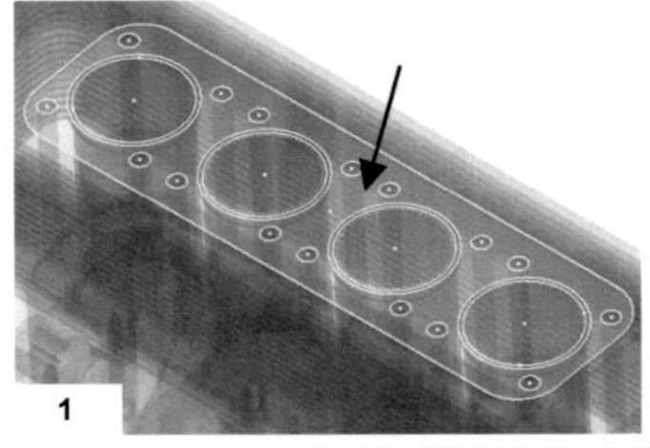

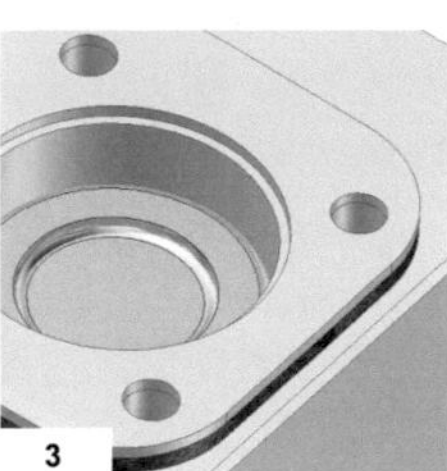

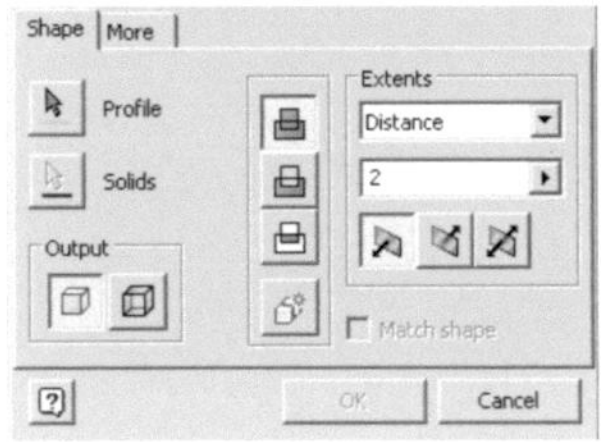

> ⬡ **Extrude**
> Profile: Marked area (picture 1)
> Operation: Join
> Extents: Distance (2)
> Direction: Shown (picture 2)
> [OK]
> ⬅ **back**

> Mark **cylinder head composition** in the model tree
> ⬡ **Color Override**
> Color: e.g. cork

Tip: Don´t extrude the area between D=47 and D= 50!

7.6.7 Inserting cylinder head, camshaft & valve

Place the components assembly cylinder head, assembly camshaft and valve into the drawing area of the assembly and assign an axial constraint of the valves to the valve guides in the cylinder head.

Make sure the valves remain axially movable.

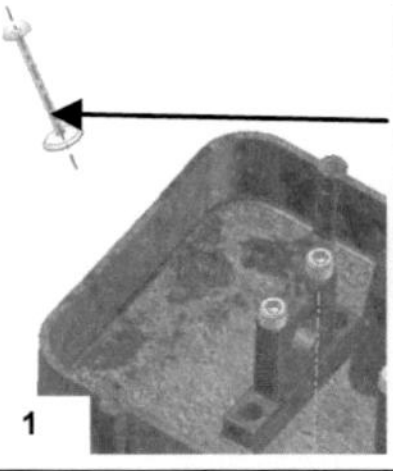

> 🖳 **Place Component**
> 1x assembly: assembly-cylinder-head.iam
> 1x assembly: assembly-camshaft.iam
> 8x part: valve.ipt

Place all components into the drawing window.

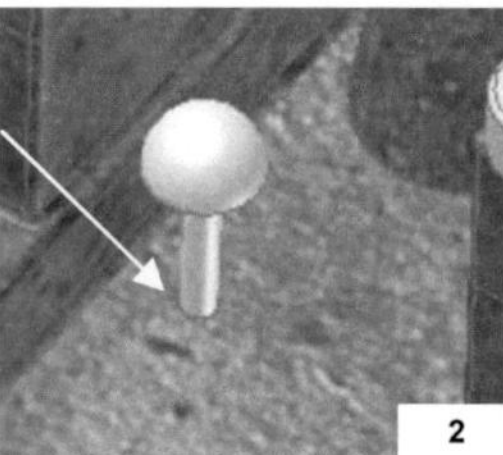

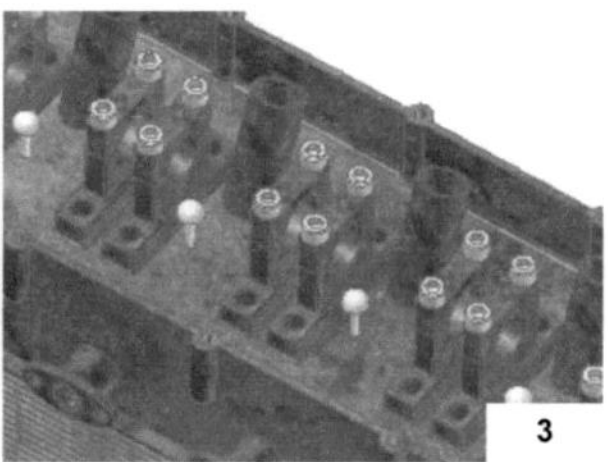

1   2   3

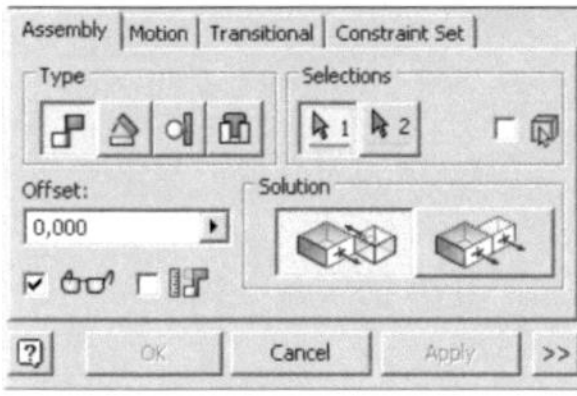

> ᵈᵖ **Constrain**
> Type: Mate
> Selections: Marked axes (picture 1+2)
> Offset: 0
> Solution: Mate
> ☐ OK

Repeat these constraints for all further 7 valves.

➢ Mark all 8 **Valves** in the model tree
➢ *farbe* **Color Override**
➢ Color: e.g. red

Tip: As all valves have to be movable heterogenic later the command 🔳 **Pattern Component** <u>cannot</u> be used here!

Then assembly camshaft is connected to the cylinder head. Use an axial constraint (picture 1) and an aerial constraint as shown in pictures 2+3.

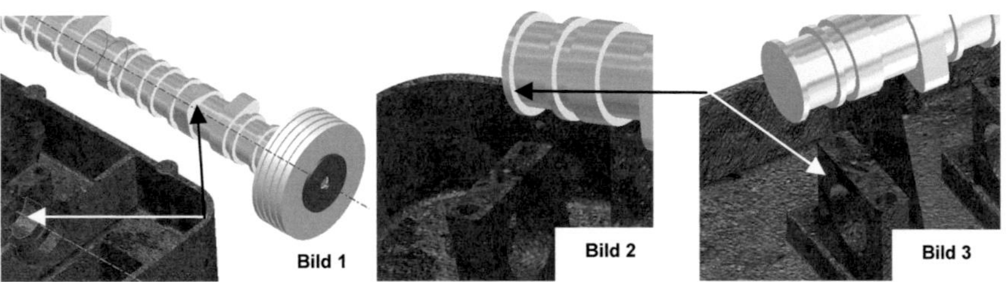

Bild 1   Bild 2   Bild 3

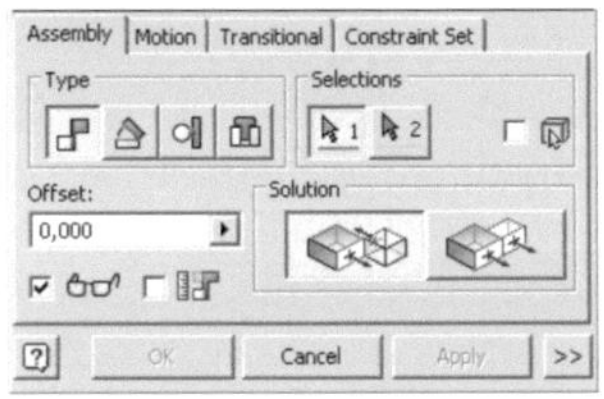

➢ 🔲 **Constrain**
➢ Type: Mate
➢ Selections: Marked axes (picture 1)
➢ Offset: 0
➢ Solution: Mate
➢ Apply
➢ Type: Mate
➢ Selections: Marked areas (picture 2+3)
➢ Offset: 0
➢ Solution: Mate
➢ OK

Place the completed cylinder head on the rest of the engine. To fixate use the seal areas (picture 1) and the axis of two hole pairs (pictures 2+3).

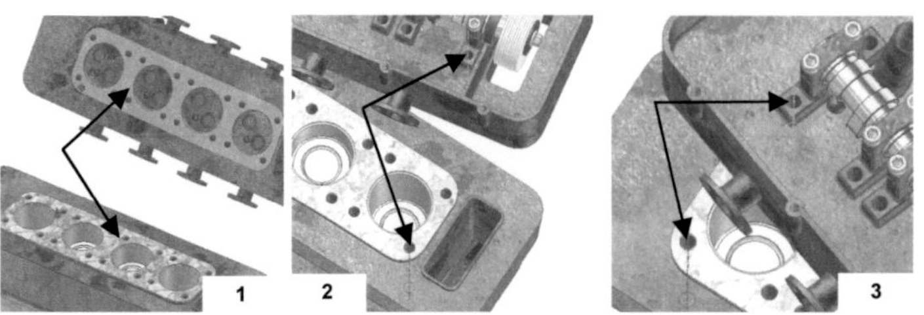

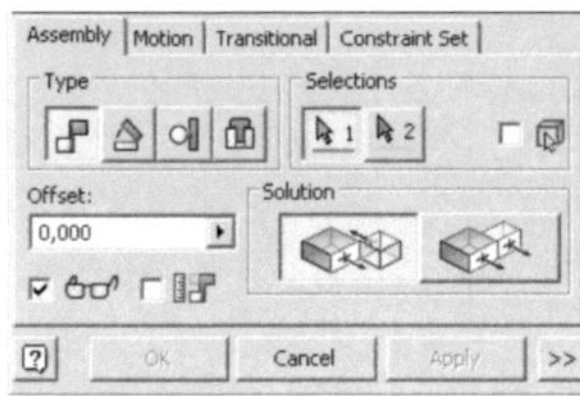

> ⬚ **_Constrain_**
> ➤ Type: Mate
> ➤ Selections: Marked areas (picture 1)
> ➤ Offset: 0
> ➤ Solution: Mate
> ➤ [Apply]
> ➤ Type: Mate
> ➤ Selections: Marked axes (picture 2)
> ➤ Offset: 0
> ➤ Solution: Mate
> ➤ [Apply]
> ➤ Type: Mate
> ➤ Selections: Marked axes (picture 3)
> ➤ Offset: 0
> ➤ Solution: Mate
> ➤ [OK]

Camshaft and valves need another constraint on each other. In this case the valves have to follow the surface guide of the cams, therefore it has to be a flexible constraint.

We use the constraint ⬚ **_Transitional_**. This means the upper stem of the valves have to follow the surfaces of the cams during movement.

The rotation movement of the camshaft will be changed into a linear movement of the valves.

It is important that the cam lobe surface (picture one) directly above the valve is chosen (when you move the mouse over the cams you can see that each cam consists of several areas).

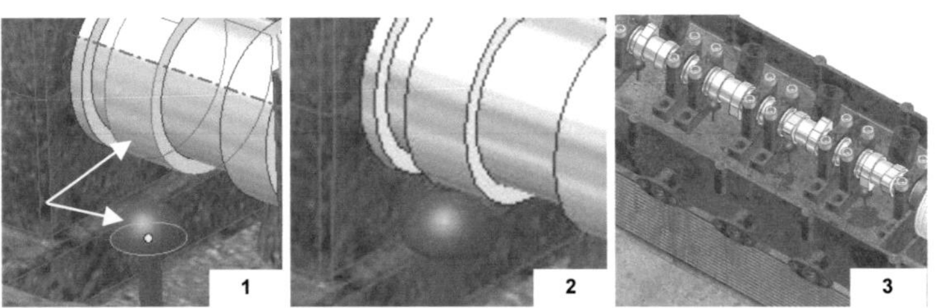

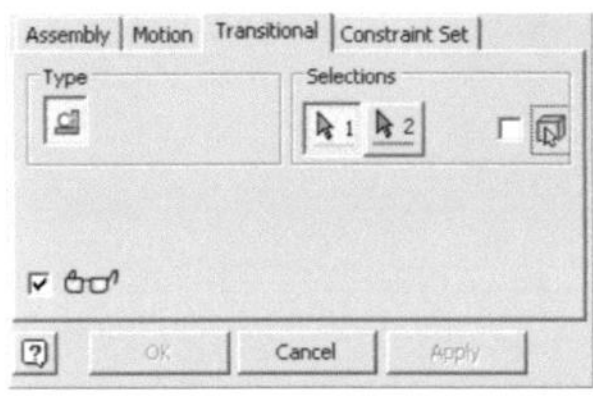

> $\Box$ **Constrain - Transitional**
> Selection 1: Valve stem
> Selection 2: Cam lobe surface (directly located one vertically above valve head)
> OK

Repeat commands for the remaining 7 valves. In this case the cams consist of several single surfaces which seamlessly adjoin each other.

Tip: It is necessary to choose the next cam lobe surface located directly above the valve, otherwise an internal computation error of the program can occur. It is also important to leave some space between valve and cam lobe surface before assigning the constraint $\Box$ **Transitional**.

Once all valves have been aligned this way, start a movement test. Click with **LMC** on a point on the camshaft and turn it.

All 8 valves now should follow linear the lobe form of the cams. Should this not be the case check if one of the valves or the cam lobe has been aligned as fixed in the model area.

7.6.8 Cylinder head screws from the content center

The cylinder head screws connect engine block and cylinder head and also are generated from the $\Box$ **Content Center**. We use Allen screws M8 x 140.

After choosing the fitting cylinder head screw choose one of the marked threaded holes (picture 1) and the contact face of the screw head (picture 2).

Then draw the screws on the red arrows to the desired length (length 140 – picture 3). Chose option $\Box$ **Place more**, to place all 16 screws at once.

> ➤ 🖨 ***Place from Content Center***
> ➤ Connective Elements – Screws – Cylinder Head.
>   Cylinder Screw with forged Allen Screw (metric)

> ➤ Click with **LMC** on the marked hole in picture 1 (program automatically chooses fitting diameter)
> ➤ Click with **LMC** on marked area in picture 2
> ➤ Drag red arrow in picture 3 with **LMC** down until you reach a screw length of 140
> ➤ Activate 📲 **Place more**
> ➤ Program automatically finds the remaining 15 threaded holes
> ➤ ☑
>
> **Note**: If you are using ***inch*** units, please transform the metric ones to inch system.

7.6.9 *Summarize identically constructed components in folders*

To reach a better overview of the components used in the assembly we can create folders in the model tree. Here useful groupings are made to simplify the tree structure. Mark the cylinder screws in the model tree and create a folder.

> ➤ Mark all 16 cylinder head screws with **LMC** (**Strg** + **LMC**)
> ➤ **RMC** > ***Add to new folder***
> ➤ Rename folder to: Cylinder Screws

Repeat these commands with the components assembly piston, valves and seals.

7.6.10　Creating a movement simulation

It is time to animate and give life to our components. The basics already have been started by a combination of the various constraints. To reach a better overview we will hide some of the components.

Mark the shown components in the model tree, then click **RMC** and remove the hook at **Visibility**. The components crankshaft and camshaft now have correctly adjusted to each other. For this create following Constraints.

 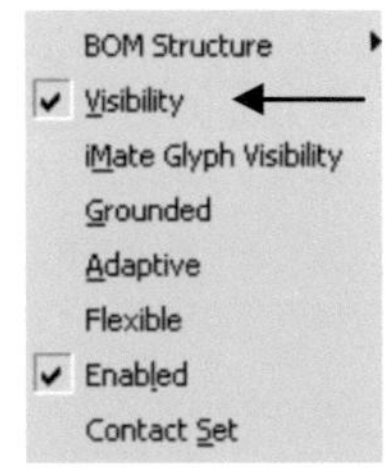

> ➢ Mark following parts in the model tree:
>
> ➢ Assembly engine block
> ➢ Assembly cylinder head
> ➢ Folder: Seals
> ➢ Folder: Cylinder Screws
> ➢ Folder: Valves
> ➢ Component order (bottom part piston rod)
>
> ➢ **RMC** > **Visibility** (remove hook)

First we position the two outer pistons upwards. For this create an Angle-constraint between Y-Z-plane of assembly crankshaft and the marked area of the engine block (picture 1).

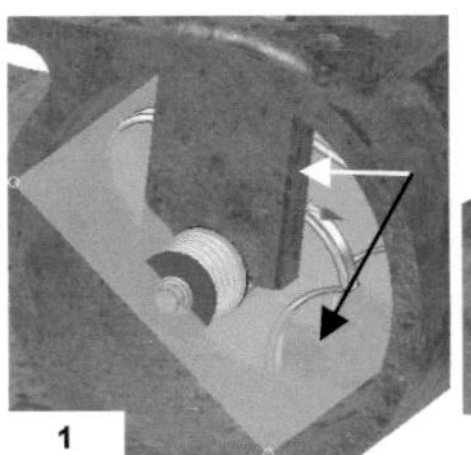

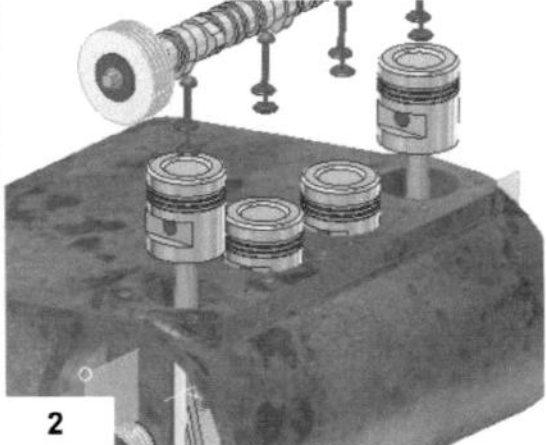

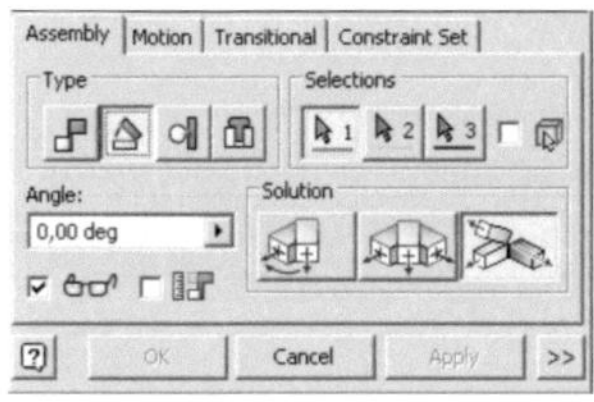

> ***Constrain***
> Type: Angle
> Selections: Y-Z-plane of assembly crankshaft and marked area of the engine block (picture 1)
> Angle: 0°
> Solution: 1
> OK

Tip: It is possible that instead of the 45° 225° have to be entered. It is important that both cams of the first cylinder point slightly downwards (picture 3).

Get the camshaft into the correct position. Create an Angle-constraint between the X-Z-plane of assembly camshaft and the marked seal area of the engine casing.

Both cams of the first cylinder have to get the position marked in picture 3.

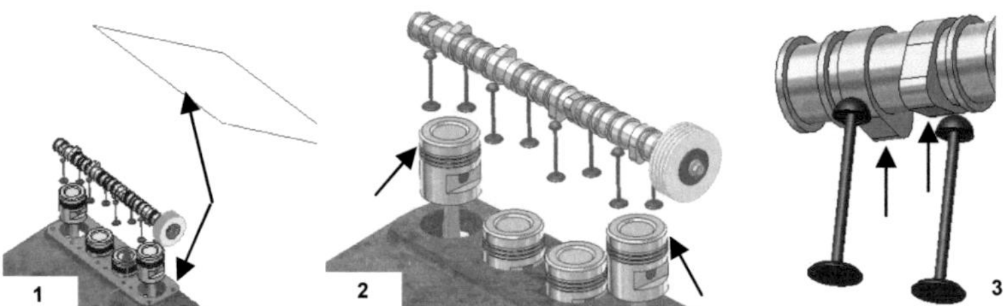

> ***Constrain***
> Type: Angle
> Selections: X-Z-plane of assembly camshaft and marked seal area of the engine casing (picture 1)
> Angle: 45°
> OK

Tip: It is possible that instead of the 45° -45° have to be entered. It is important that both cams of the first cylinder point slightly downwards (picture 3).

Our engine works according to a four-stroke principle (suction, compress, combust and exhaust).

1. Stroke: Suction

The piston moves down. The exhaust valve is closed and the intake valve opened. During the downwards movement of the piston the air/fuel mixture is drawn into the cylinder.

2. Stroke: Compress

The piston moves up. The air/fuel mixture is compressed. Both valves are closing.

3. Stroke: Combust

After the piston top dead center (TDC) ignition occurs. The air/fuel mixture explodes and the piston moves down. Shortly before the piston reaches its bottom dead center (BDC) the exhaust valve opens.

4. Stroke: Exhaust

The exhaust valve is open. The piston moves up and moves the burned air/fuel mixture from the combustion chamber.

The rotary ratio between crankshaft and camshaft has to be exactly 2:1. Both pulleys are already created in the proper size, however, for the simulation we have to state this ratio again.

For a better depiction you can remove the visibility for the component engine casing. Then create a new movement constraint.

It is not necessary that both components have a mechanical contact to each other.

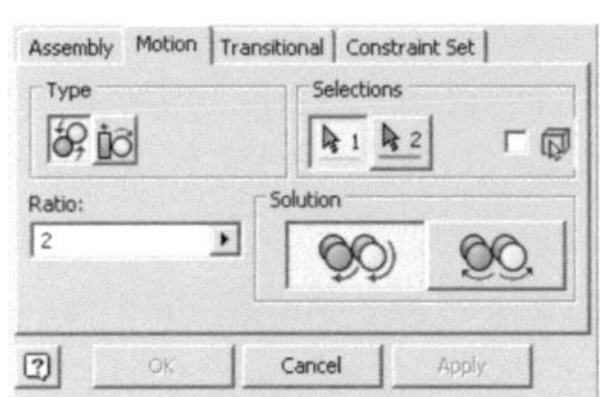

> ⌐ **Constrain**
> Type: Motion – Rotary movement
> Selection 1: Pulley camshaft (picture 1)
> Selection 2: Pulley crankshaft (picture 2)
> Ratio: 2
>   OK

7.6.11 Simulation of the components & creation of a video

Before we can animate the crankshaft and camshaft we have to remove the just created angle constraint of the camshaft again (you will find it in the model tree under the component assembly camshaft at last place).

It was simply meant to correctly tweak the rotation position of camshaft to crankshaft. Under assembly crankshaft (180°) choose the angle constraint with **RMC** in the model tree (you will find it under the component assembly crankshaft at last place) and select the command **RMC > Drive Constraint**.

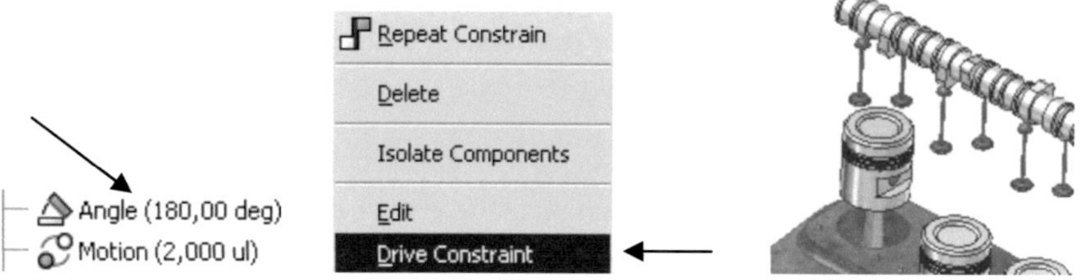

> **RMC > Drive Constraint** (on angle constraint of assembly crankshaft)
> Start: 0° / 180° (depending on pre-setting)
> End: 720°

> ⊙ **Start Recording**
> File type: *.AVI
> File name: Video
> Compression: Microsoft Video 1
> ▶ **Start Simulation**
> Wait until the simulation has ended
> ⊙ **End Recording**
> OK

You now can view the video at the saving location.

Tip: When choosing the format *.wmv you can add further settings in regards to the quality of the video.

7.6.12   Part timing belt

The movement simulation already has been created. To also join the components camshaft and crankshaft mechanically we create the part **timing-belt.ipt** as connective component.

Create a new part on the Y-Z-plane of the crankshaft, then draw the shown outline (project edges of the pulley of the camshaft) and then create a second ✍ **2D Sketch** for the path of a Sweep-object.

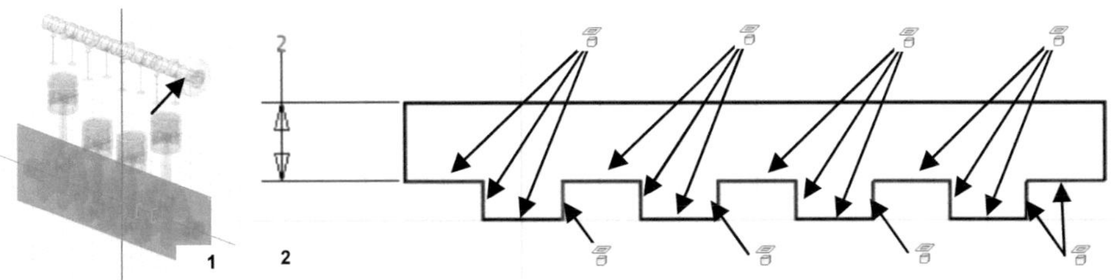

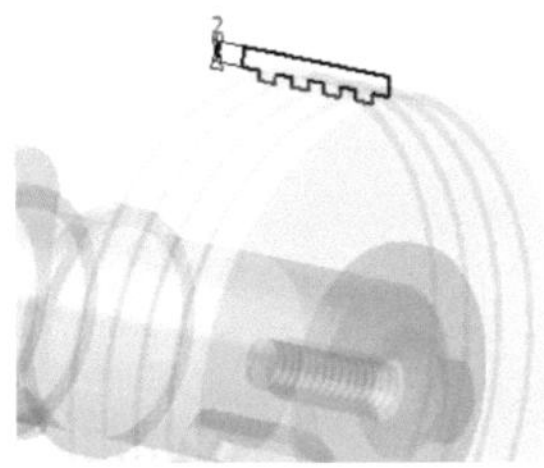

> 🗇 **Create**
> Name: timing-belt.ipt
> Position: Y-Z-plane of the crankshaft (picture 1)
> 🖥 **Project Geometry** (marked edges of the camshaft pulley)
> Draw shown outline (picture 2)
> ✔ **Finish Sketch**

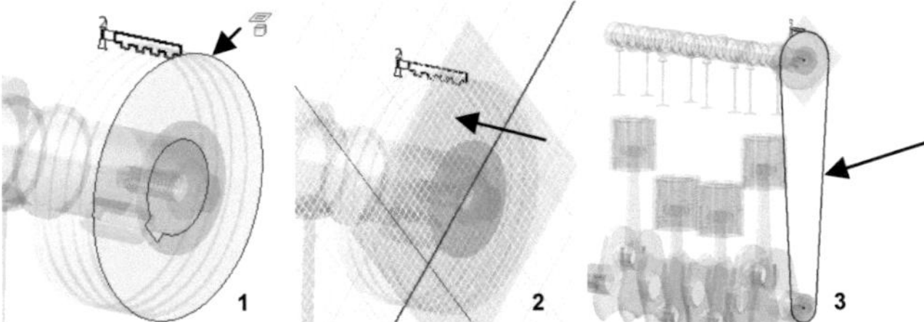

> 🗎 **2D Sketch** (on marked area picture 2)
> 🖥 **Project Geometry** (outer edges of camshaft-pulley and crankshaft-pulley picture 4)
> Define both as ⊥ **Construction Lines**
> Draw the 2 tangents on the projected circles
> Draw the 2 circular arcs (marked picture 4), concentirc to the projected circles, ending on the tangents
> ✔ **Finish Sketch**

Tip: Make sure to use the respective outer edges of the pulleys. First project the outer circles, then define them as ⊥ **Construction Lines** and place the tangents on it. Then draw two circular arcs to get a closed outline for the sweeping-path.

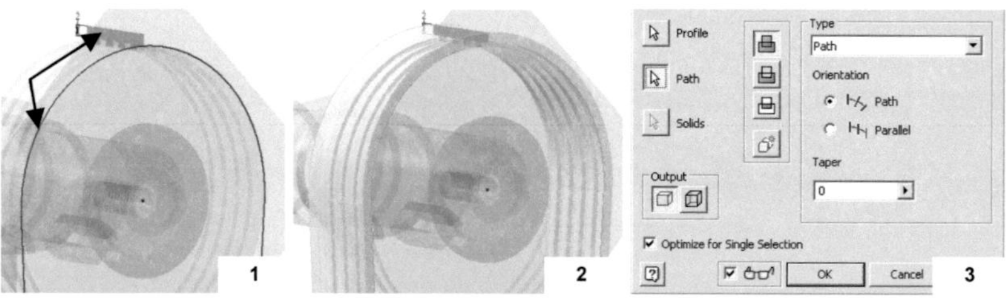

> ⏹ **Sweep**
> Profile: Sketch 1
> Path: Sketch 2
> ⏹ OK

> ⏹ **Back**

> Select **timing belt** in the model tree
> ⏹ **Color Override**
> Color: e.g. rubber (black)

> ⏹ **Save > Save all**

7.6.13   Valve covers, valve cap composition, washers & bolts

The following components (valve cover, valve cap composition, washer: DIN 125-A 4.3 and bolt: DIN 6912 – M4 x 50) you load from the 'download file' and then place them into the assembly.

Then use the command ⏹ **Constrain**, to position the components as shown.

  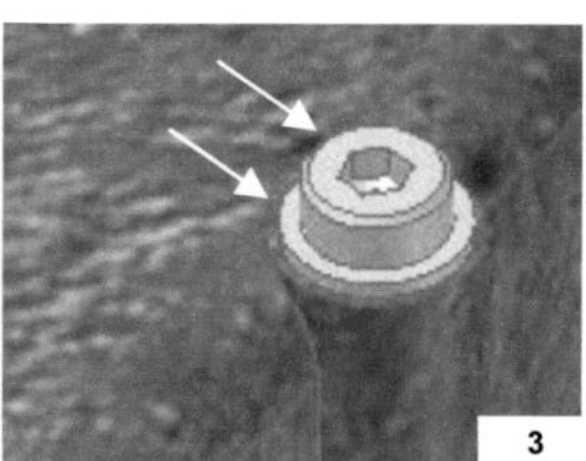

> 🖨 ***Place Component***
> ➤ 1x Part: valve-cap.ipt (new part from download)
> ➤ 1x Part: valve-cap-composition.ipt (download)
> ➤ 10x Washers: DIN 125 - A 4.3.ipt (download)
> ➤ 10x Bolts: DIN 6912 - M4 x 50.ipt (download)

Order all components according to the three pictures.

7.6.14   Imprint valve cover

We still want to add an imprint to our valve cover. For this we *edit* the part in the assembly, create a new ✎ ***2D Sketch*** on the topside of the valve cover and there create the shown **A** ***Text***.

Then we use the command ⟡ ***Emboss*** to imprint the text 1 deep into the part and then return to the main assembly with ↩ ***Back***.

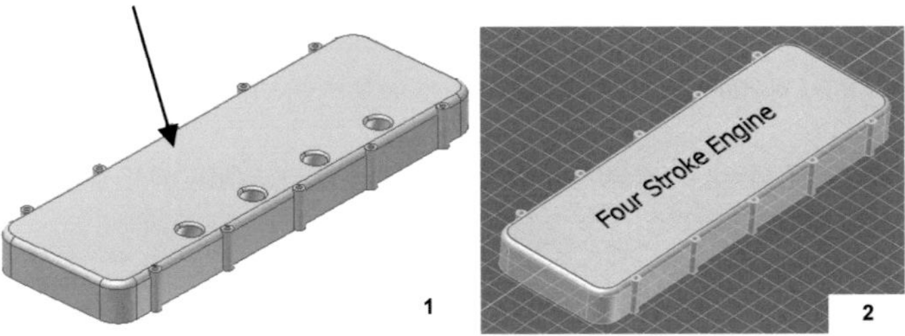

> ➤ ***RMC*** on ***valve cover*** > ***Edit***
> ➤ ✎ ***2D Sketch*** (on marked area picture 1)
> ➤ **A** ***Text***
> ➤ Size: 20
> ➤ Text: Type in ***Four Stroke Engine***
> ➤ [OK]
> ➤ ✔ ***Finish Sketch***

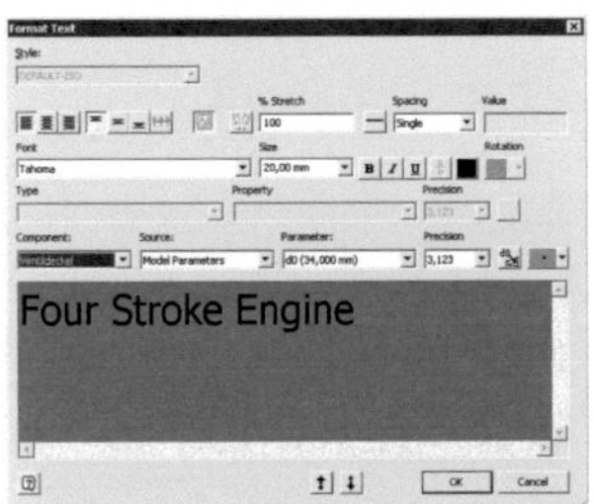

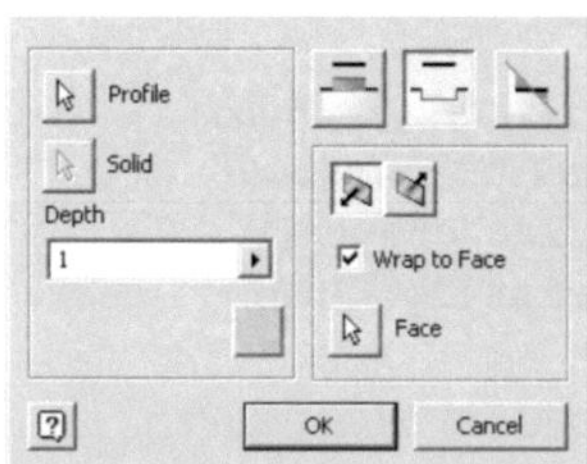

> ➤ 🖼 *Emboss*
> ➤ Profile: Sketch with text
> ➤ Depth: 1
> ➤ Procure on area
> ➤ Area: Marked area (picture 2)
> ➤ [OK]
>
> ➤ ◄ *Back*
> ➤ 🖬 *Save*

assembly-four-stroke-engine.iam now is complete. 🖬 *Save* and ✖ *Close* the file.

8   *Exercises in the area CREATING DRAWINGS*

8.1   *Drawing creation part camshaft-pulley*
8.1.1   *Editing of Sheets & iProperties*

In our first example of drawing creation we will use the part *camshaft-pulley.ipt*. Create a 🖼 *New Drawing* and 🖬 *Save* it as *drawing-camshaft-pulley.idw* in the project folder.

To give our drawing sheet the desired size, choose the command *Edit Sheet* with *RMC* in the model tree on *Sheet:1*.

The window *Edit Sheet* opens. Here you choose the two settings *A4* and *Horizontal* format. Then we change a few settings in the area *iProperties*.

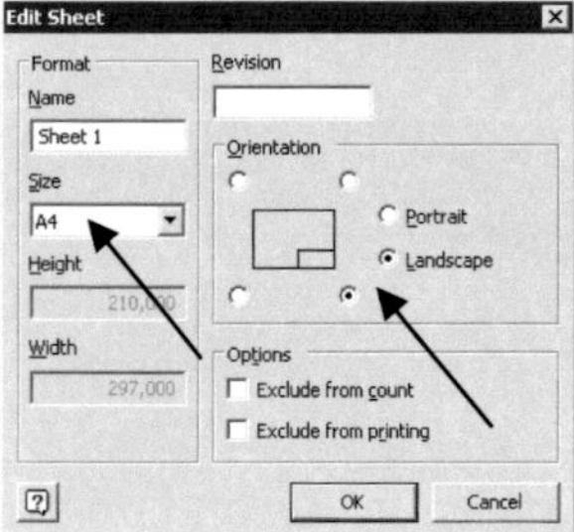

> ➤ *RMC* on *Sheet:1* in the model tree *Edit Sheet*
> ➤ Size: A4
> ➤ Orientation: Landscape
> ➤ [OK]
>
> ➤ *RMC* on *drawing-camshaft-pulley* in the model tree *iProperties*
> ➤ Tab: Project
> ➤ Part Number: Camshaft-Pulley
> ➤ [OK]

8.1.2 Creation of basic and parallel views

Now we can create the first ▦ **Base** view. Select the file **crankshaft-pulley.ipt** and, after adding the stated basic settings, place it in the drawing window. Then create two ▦ **Projected** views. Note that you can create parallel views until you close it with the command **RMC > Done**.

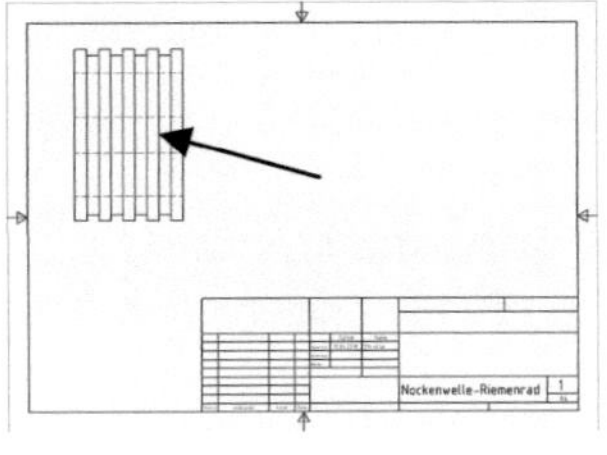

> ▦ **Base**
> ➤ Tab: Component
> ➤ File: In project folder **crankshaft-pulley.ipt**
> ➤ Scale: 3:1
> ➤ Alignment: Front
> ➤ Style: With covered lines
> ➤ Place view in sheet (**LMC**)
> ➤ **ESC**

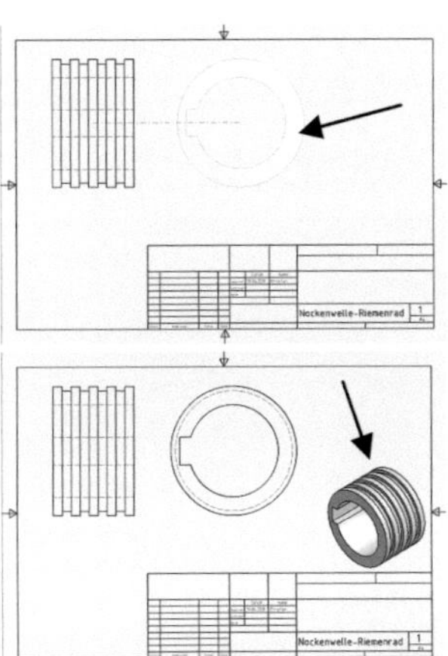

> ▦ **Projected**
> ➤ Select base view
> ➤ Drag mouse to the right
> ➤ Place projected view 1
> ➤ Drag mouse to the bottom right
> ➤ Place projected view 2
> ➤ **RMC > Create**
>
> ➤ **RMC > Edit View** on isometric view
> ➤ Style: ▦ Shaded
> ➤ [OK]

Tip: Already placed views can be moved when you mark the view with **LMC** and then move along the appearing <u>red</u> rimmed border.

8.1.3 Creating & retrieving of dimensions

The views are created, now we can start dimensioning them. Here the program offers good help.

With the command ▦ **Retrieve** (**RMC** on the respective view) already available dimensions can be retrieved and shown. Do this with the first view. Then create a new dimension via manual dimensioning.

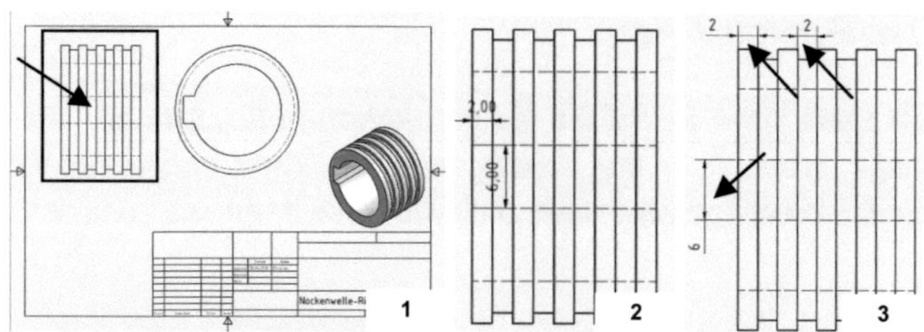

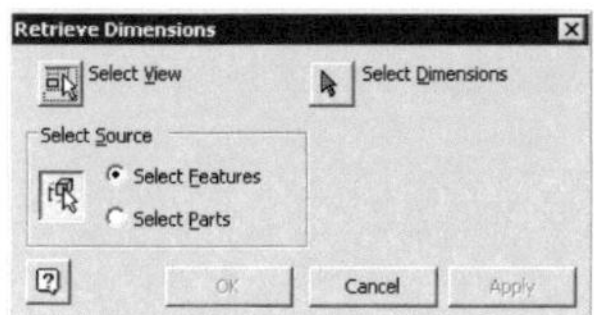

> ➤ **RMC** on the left basic view
> ➤ 🖺 **Retrieve Dimensions**
> ➤ View: Drag window with **LMC** over view 1 (picture 1)
> ➤ Dimensions: Choose three marked dimensions (pic. 3)
> ➤ [OK]

Tip: To move the dimensions you can either drag the single dimensions with **LMC** to the desired position or use the command 🗔 **Arrange** to create an automatic order of the dimensions.

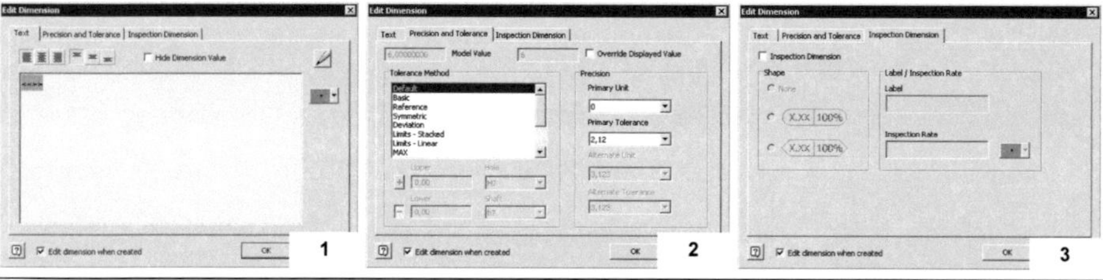

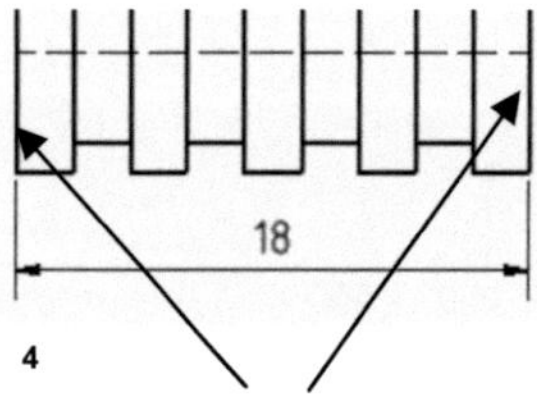

> ➤ 🗂 **Dimension**
> ➤ First the left, then the right edge (marked)
> ➤ Then place dimension with **LMC**
> ➤ Tab: Precision and Tolerance
> ➤ Tolerance Method: Standard
> ➤ Precision: Main unit, and -tolerance: 0
> ➤ [OK]

Dimension the shown view and use the command ┼ **Center Mark**, to mark the center of the rotary part. Click the lines and draw the ends on the green points into the desired length.

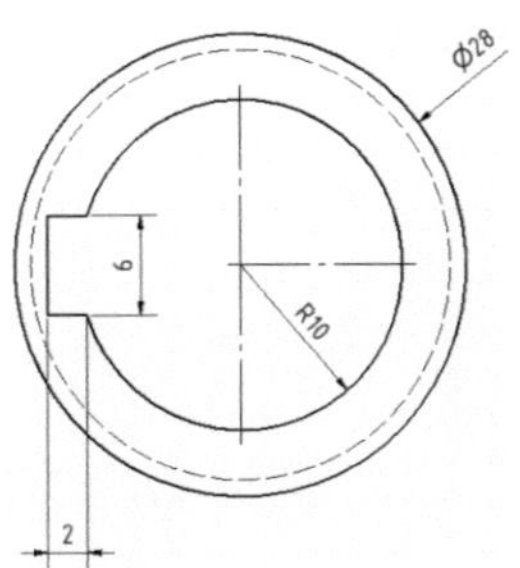

> ⊓ *Dimension*
> Insert dimensions 2, 6, R1= 10 and D2= 28 manually

> ✛ *Center Mark*
> Select one of the two circle lines, place the center point marking and then change the length of the lines (drag and drop on the end points)

camshaft-pulley.idw now is complete. 🖫 *Save* and ✕ *Close* the file.

8.2 Drawing creation part PISTON
8.2.1 Creation of cross-sectional views

In the following exercise we will create a ⊟ *Section*.

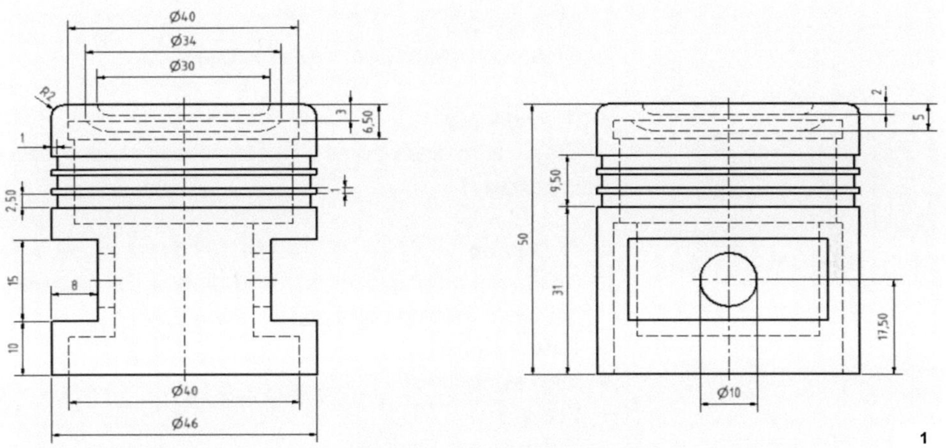

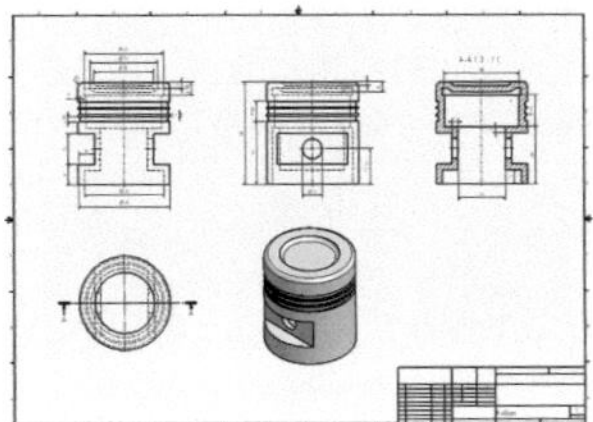

Create a 🗋 *New Drawing* and 🖫 *Save* it as *drawing-piston.idw* in the project folder.

Create the shown 🖼 *Basic-* and 🖼 *Projected* views of the piston as well as a ⊟ *Section* and dimension it.

Make sure not to double dimension and use 🗒 *Retrieve Dimensions* as well as manual dimentioning.

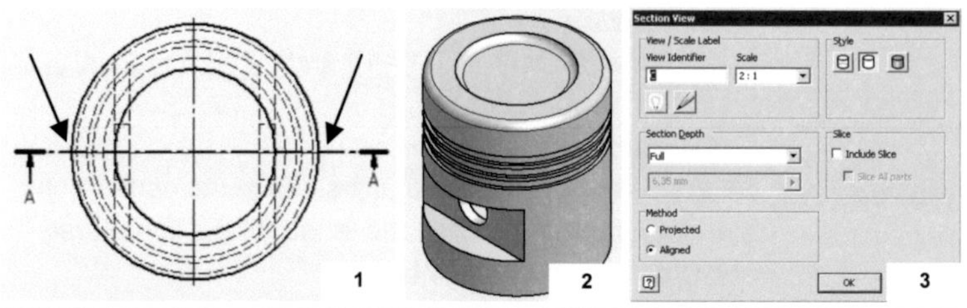

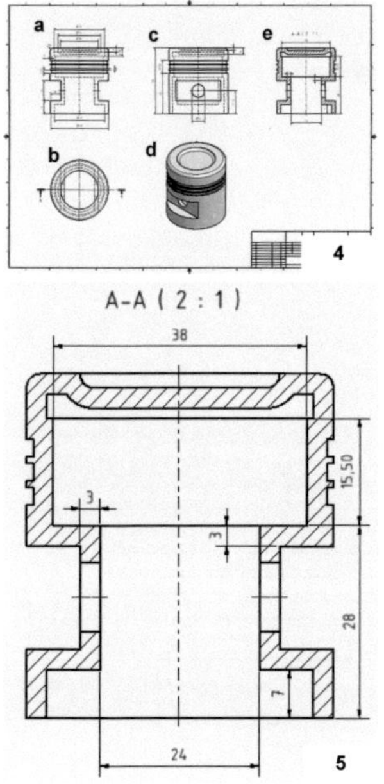

> 🖼 **Base**
> Tab: Component
> File: in project folder *piston.ipt*
> Scale: 2:1
> Alignement: Front
> Style: With covered lines
> Place view in sheet (**LMC**) (picture 4a)
> **ESC**

> 🖼 **Projected**
> Create the two parallel views (picture 4b,c)

> 🖼 **Projected**
> Create a parallel view as isometric view in shaded view (picture 4d)

> 🖼 **Section**
> Select top view (picture 4b) and draw a line between the outer, marked circle points
> **RMC** > **Done**
> View symbol: A
> Scale: 2:1
> Depth of cut: Complete
> ☐ OK

Tip: You can edit all dimension arrows, lines, shadings and views when you mark them and then click with **RMC** on **Edit**.

Place the cross-sectional view as shown in picture 4e into the drawing area. Sufficiently dimension all views. **piston.ipw** now is complete. 🖫 **Save** and ✕ **Close** the file.

8.3 Drawing creation Assembly PISTON
8.3.1 Position numbers

In the following exercise we will use the assembly *piston.iam*. We exercise the creation of position numbers, parts lists, detail drawings and segments.

Create a ▦ *New Drawing* and ▤ *Save* it as *drawing-assembly-piston.ipw* in the project folder.

Insert the separate 6 parts of the assembly piston as respective ▦ *Base* views and order them as shown. Then create an isometric as well as shaded ▦ *Base* view of *assembly piston* and provide them with ⊕ *Auto Balloon*. Then the numbers are manually ordered (drag and drop).

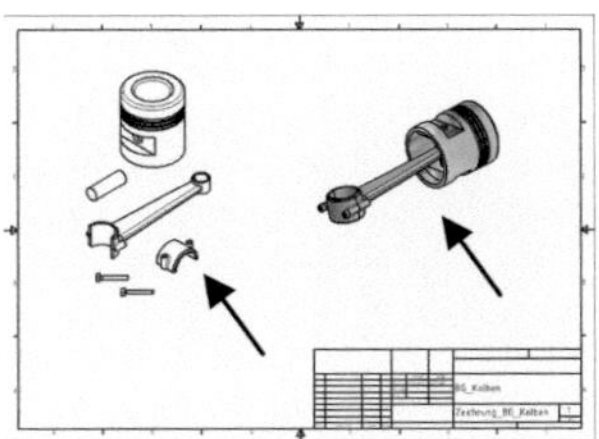

> ▦ *Base*
> Create an isometric, shaded basic view in the assembly *assembly-piston.iam*

> ▦ *Base*
> Create 6 isometric basic views in *assembly-piston.iam* contained components (piston, bolt, connecting rod topside- and bottom side and both screws)

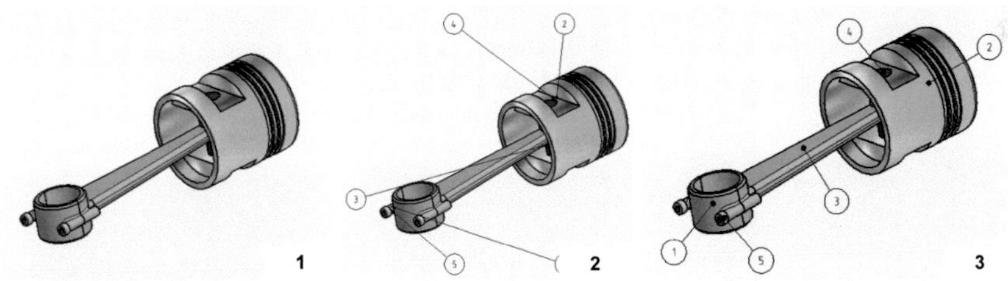

> ⊕ *Auto Balloon*
> Select View Set: Isometric view of *assembly piston*
> BOM View: structured
> Placement: Horizontal, Offset Spacing: 0
> Balloon Shape: Round
> OK

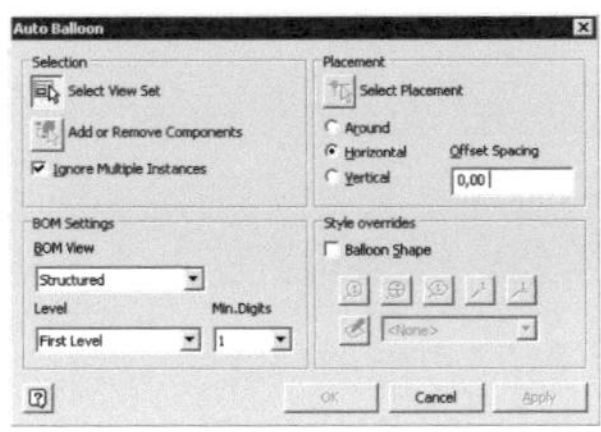

Tip: To position the position numbers in the desired order click them with **LMC** and move them (drag and drop).

8.3.2   Creating & Editing of Parts Lists

After the position numbers have been created we create a ▤ **Parts List** for our assembly.

Place the parts list into the drawing window and then edit it (**RMC** > **Edit Parts List**). Complement the missing data in the area **part number** and **description** and sort the parts then according to **Object Number**.

Then use the command ⊙ **Balloon** to also number the basic views of the single parts.

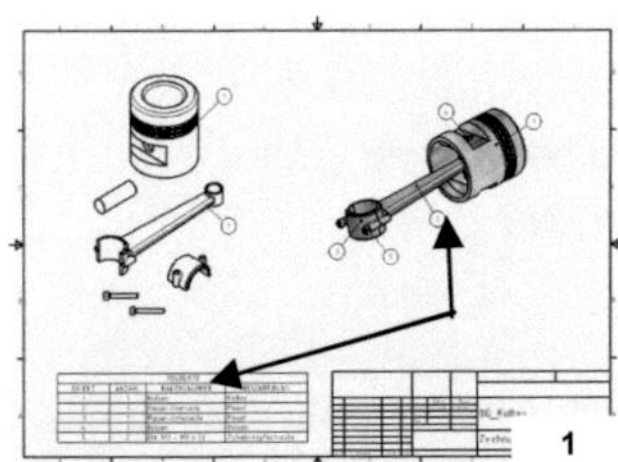

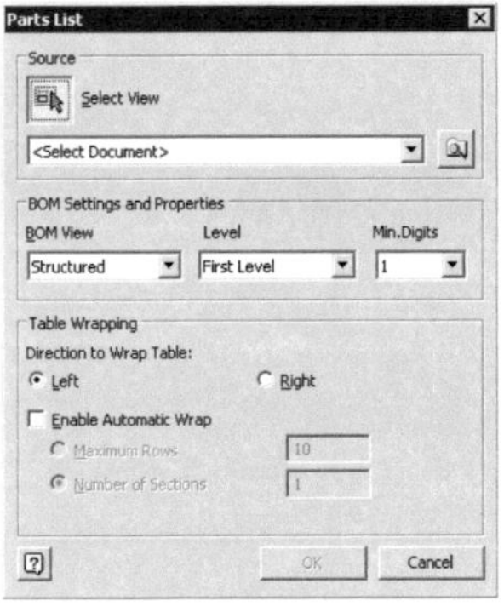

> ▤ **Part List**
> View: Click on marked view (picture 1)
> [OK]

> Place table in drawing
> **RMC** on table: Edit parts list

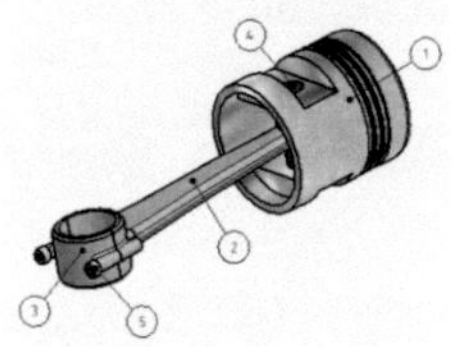

> Edit the columns **part number** and **description** as shown
> ⇵ **Sort** the parts according to **object number**
> [OK]

Tip: The numberings of the already set position numbers are refreshed automatically.

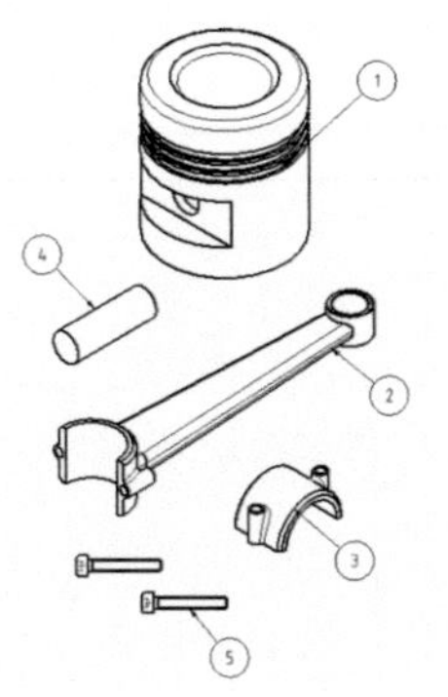

> ① **Balloon**
> Select piston
> Place position numbers
> **RMC**: Next
> Repeat this step with the remaining parts
> **RMC** on first created position number: **Edit Position Number**
> Assign each position number the correct value this way (see table above)

8.3.3 Create detail drawings

Now we will create a detail drawing. In our example we use the connecting rod end to better show the chamfer. Use the command ⓦ **Detail** and follow the command chain.

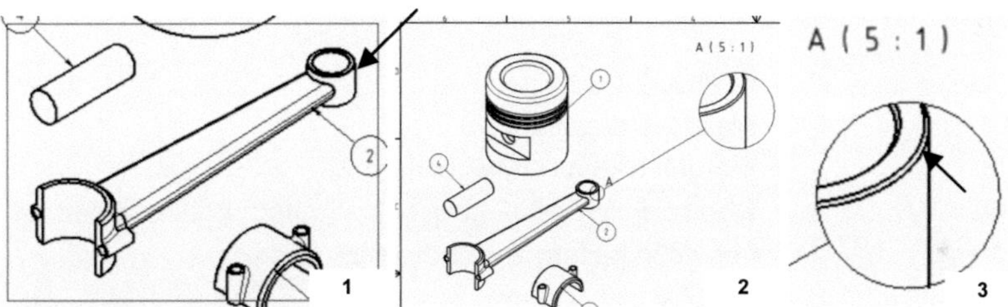

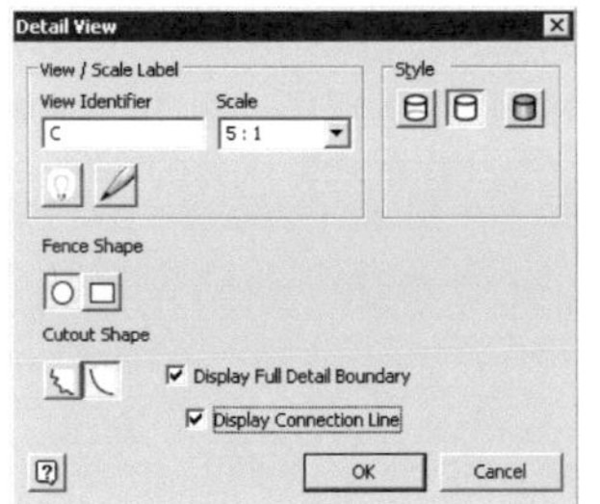

> ⓦ **Detail**
> Mark the view Connecting rod end - topside
> View Symbol: A
> Scale: 5:1
> Style: Without covered lines
> Form of the Detail Frame: Round
> Segment-form: Line, set hook 2x
> OK

8.3.4 Creation of break outs

A break out is created similar to a detail drawing. However, here it is necessary to first create a new ✎ **2D Sketch**. For this it is important to first mark the view to be sliced with **LMC** (sketch and view otherwise are not connected). Create a new base view of the piston (shaded), then draw a circle (D= 70) and end the sketch again. Use the command ⮊ **Break Out** and follow the commands.

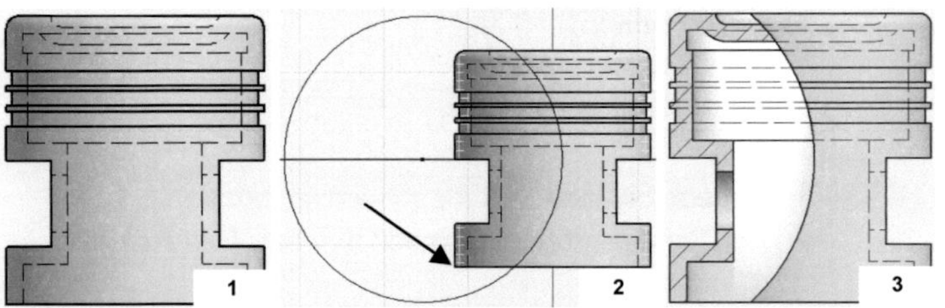

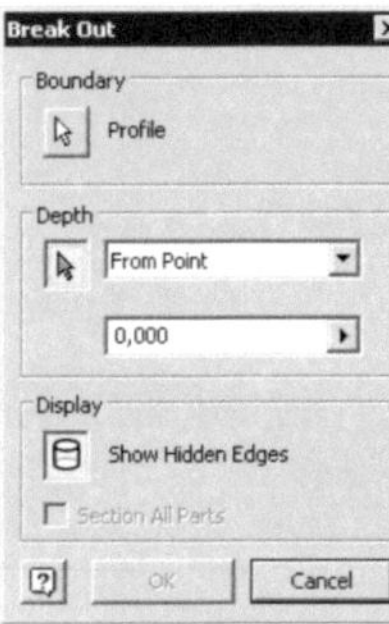

- ➢ 🖻 **Base**
- ➢ New base view of the piston (picture 1)
- ➢ Position the view with **LMC**
- ➢ **ESC**
- ➢ Mark this view with **LMC** (a red border must be visible)

- ➢ 🖉 **2D Sketch**
- ➢ Draw a circle D= 70 as shown (picture 2)
- ➢ ✔ **Finish Sketch**

- ➢ 🖻 **Break Out**
- ➢ Mark the created view
- ➢ Margin: Sketch circle (picture 2)
- ➢ Depth: From point (mark the outer, bottom point as shown in picture 2 with distance 0)
- ➢ OK

> **Tip**: With the selection **Depth** you determine the depth of the cut and its reference. By choosing the left corner point of the piston in our example we reference to the cusp of the piston edge. Also you could use the center point of a bolt hole or any other reference.

The interaction of reference and depth is decisive. ***assembly-piston.idw*** now is complete. 🖬 ***Save*** ✖ and ***Close*** the file.

9   *Exercise PRESENTATION*

Create a 🖻 ***New Presentation*** for our example and 🖬 ***Save*** it as ***presentation-assembly-four-stroke-engine.ipn*** in the project folder. Then create a new 🖻 ***View*** and use the explosion method ***Automatic*** with a distance of 500.

This method is very vague and often requires a lot of editing as the single paths still have to be edited separately. However, for our exercise it is sufficient.

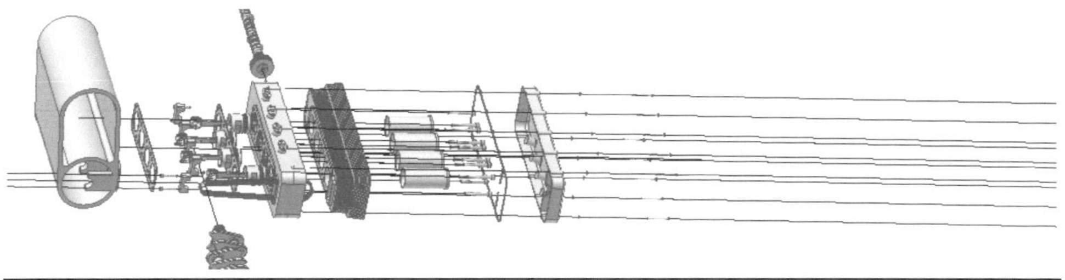

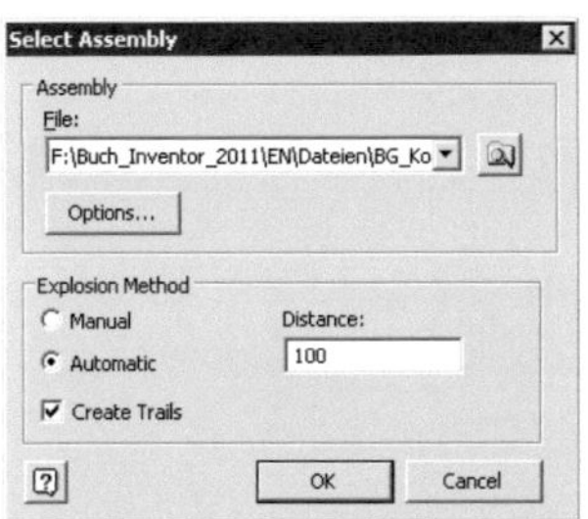

> 🖳 ***Create View***
> ➢ File: assembly-four-stroke-engine.iam
> ➢ Explosion Method: Automatic
> ➢ Distance: 500
> ➢ [OK]

Tip: A manual explosion method is a lot more exact and from the start you can determine separate paths and distances of the components.

To be able to change the paths use the command ⌖ ***Tweak Components***. Here you can rotate or move components and therefore get them into the desired position. When all component positions are according to your perceptions use the command ▶ ***Start*** to begin the presentation.

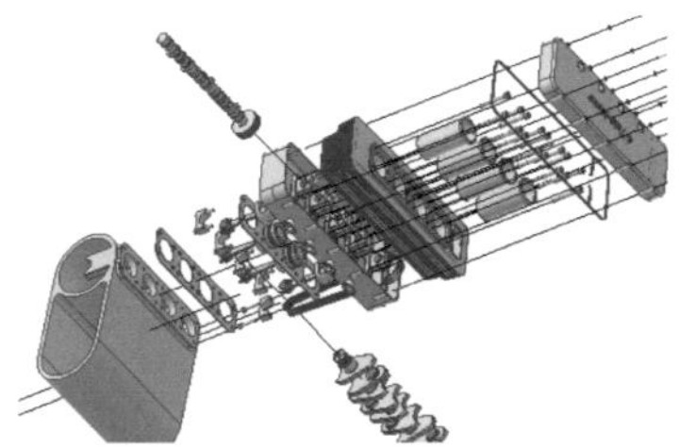

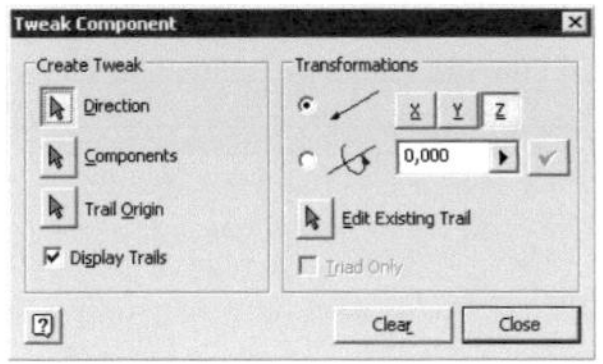

> ➢ ⌖ ***Tweak Components***
> ➢ Direction: Place moving vectors
> ➢ Components: Single components
> ➢ Transformations: Direction (X, Y, Z) or Rotary angle
> ➢ Move components
> ➢ [Close]

Tip: You can change existing paths as follows:

> ⬚ *Tweak Components* > *Edit Existing Trail*

A precise rotation of the component can be done with the command ⊗ *Precise View Rotation*.

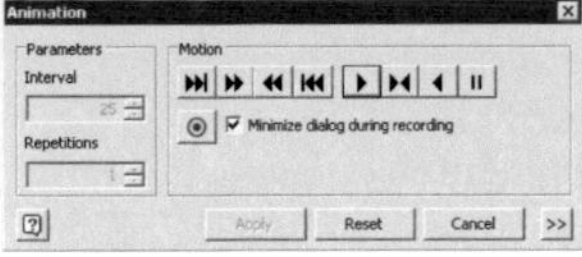

> *Parameters*: Sets the number of intervals and their repetitions
> ▶ *Start*: Begins the presentation
> ◉ *Video*: Save the animation as a video file

presentation-assembly-four-stroke-engine.ipn now is complete. 🖫 *Save* and ✕ *Close* the file.

10 Exercise INVENTOR STUDIO

10.1 RENDERING of the assembly FOUR-STROKE-ENGINE

As a small example we will render the entire four-stroke-engine. Open ***assembly-four-stroke-engine.iam*** and start the command ⬭ *Render Image.*

> Tab: *Environments* > ⬭ *Inventor Studio* > ⬭ *Render Image*

Set maximum quality and begin. Then 🖫 *Save* the picture in your project folder.

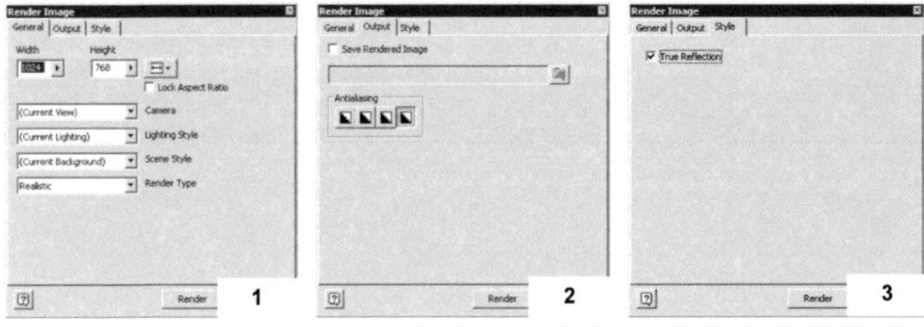

> ➤ 📂 **Open** (assembly-four-stroke-engine.iam)

> ➤ Tab: **Environments**
> ➤ 🔦 **Inventor Studio**
> ➤ 🍵 **Render Image**
> ➤ Tab: General
> ➤ Width / Height: 1024/ 768
> ➤ Tab: Issue
> ➤ Antialias: Tops
> ➤ Set saving location
> ➤ Render

Switch to desktop and open your project folder. Here you will find the generated picture.

11 Exercise for SHEET METAL CREATION

The area **sheet metal part creation** basically is similar to the area **part creation**. Here you will find special commands for the creation of sheet metal parts. For our example we will construct an oil pan which later can be placed under the engine.

Create a new 📄 **SheetMetal** part and 💾 **Save** it as **oilpan.ipt** in your project folder. Draw the shown rectangle and close the sketch. Then edit the 🖫 **Sheet Metal Defaults**.

11.1 Creating the oil pan

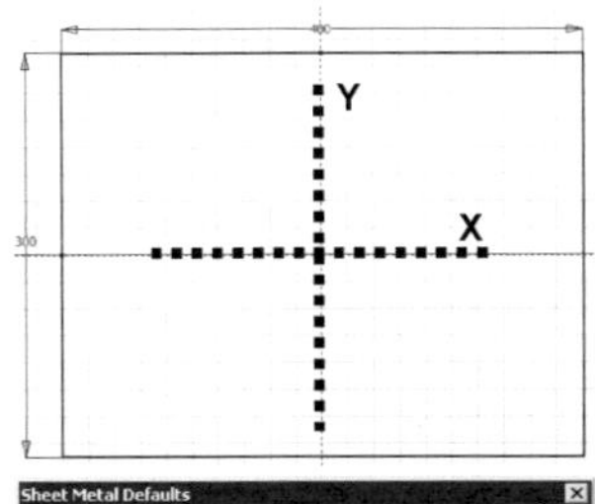

> ➤ 🗋 **New**
> ➤ 📄 **SheetMetal.ipt**
> ➤ 💾 **Save** (oilpan.ipt)
> ➤ 🖫 **Project Geometry** (X-Y-Z-axes)
> ➤ Rectangle: draw rectangle 400 x 300 and align symmetrically on the axis
> ➤ ✓ **Finish Sketch**

> ➤ 🖫 **Sheet Metal Defaults**
> ➤ Adopt thickness from rule: Remove hook
> ➤ Thickness: 1
> ➤ OK

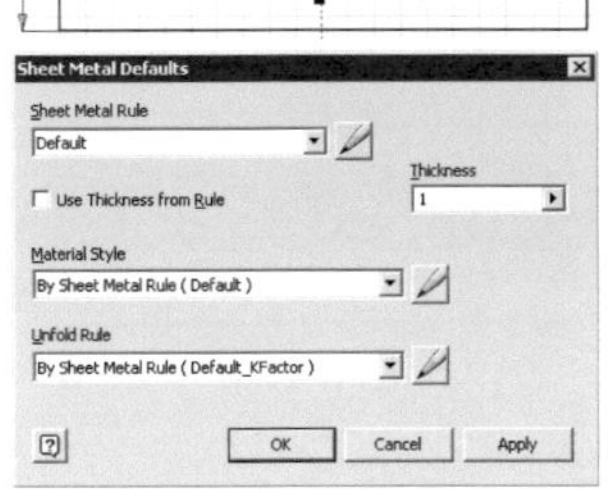

Tip: In the sheet *metal standards*: *sheet metal rules*, *material styles* and *execution rules* are defined. You can use the given standard values or assign your own. Further setting options you find under *Style Editor* (*Manage* > *Style-Editor*).

Create a sheet with the command ☐ *Face*. Then we create a ⊲ *Flange* on each of the 4 marked edges.

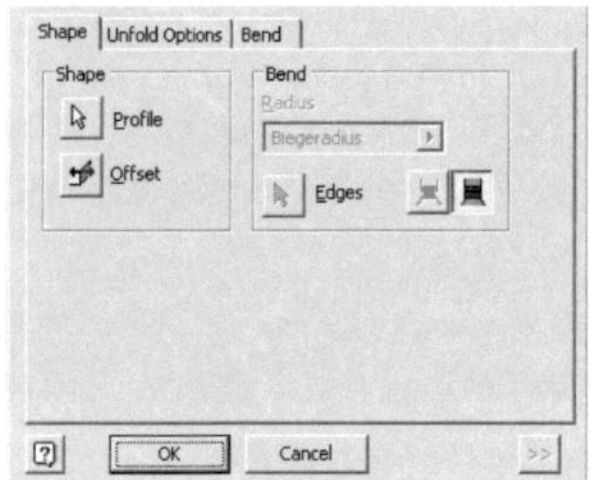

> ☐ *Face*
> Tab: Shape
> Profile: Rectangle
> Incur all pre-settings
> ▭ OK

11.2 Creating the flaps

Repeat the command ⊲ *Flange* to create the marked grindings on the newly created sheet metal pieces. Use the command 🗐 *Unfold*, to generate a sheet execution and then the command 🗐 *Create Flat Pattern*, to go back.

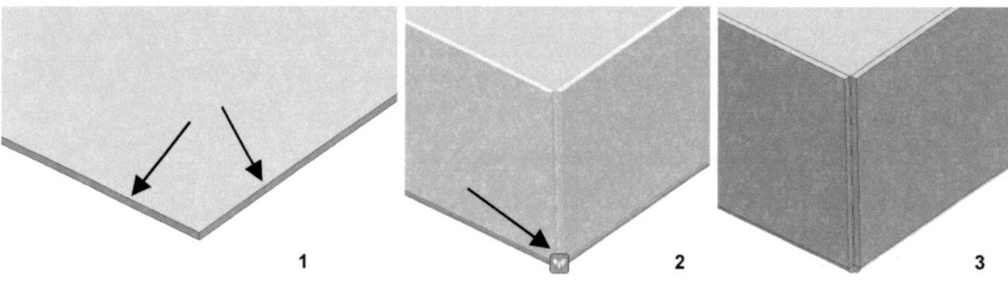

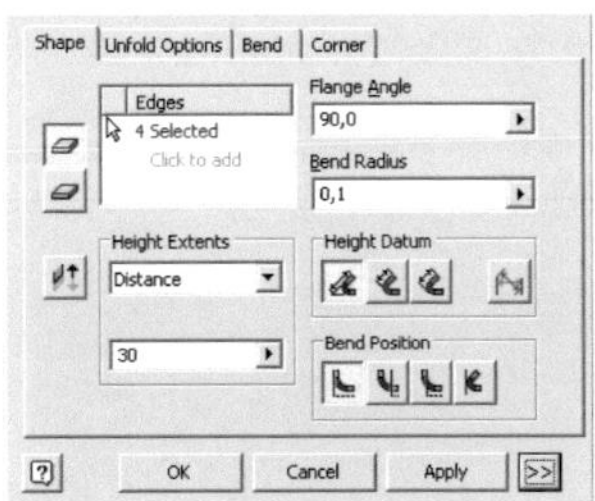

> ↦ *Flange*
> Tab: Shape
> 4 Edges of the upper area
> Height Extents: Distance (30)
> Flange Angle: 90°
> Bend Radius: 0.1
> Height Datum: First
> Bend Position: First
> OK

Tip: With the command ✉ *Corner Seam* a corner already can be defined (e.g. distance of the sheet edges to each other).

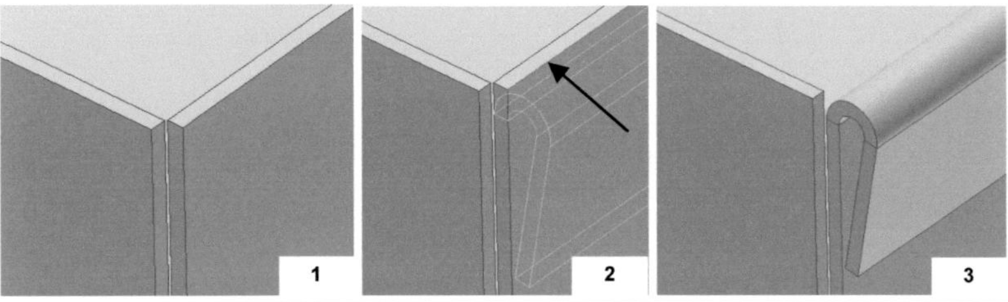

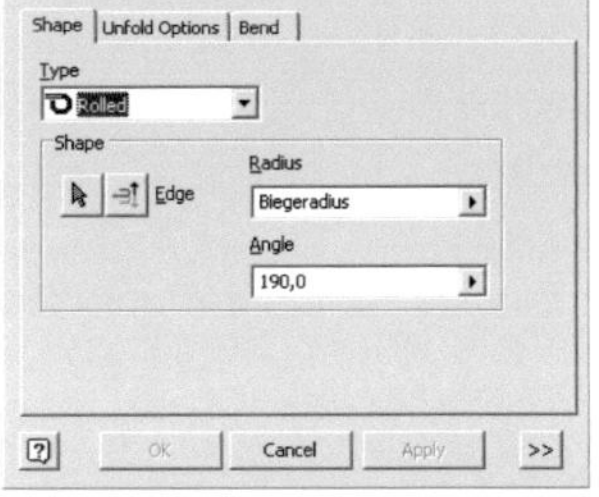

> ↦ *Flange*
> Tab: Shape
> Type: Rolled
> Shape: Marked edge (picture 2)
> Radius: Bending Radius
> Angle: 190°
> OK

Repeat the commands for the remaining 3 sheet sides. 💾 *Save* and ✖ *Close* the file.

12 Creating a WELD CONSTRUCTION

The area of the welding construction assembly is basically similar to the area assembly creation. Here you will find special commands for the creation of welding construction assemblies. As a short demonstration we will start a small exercise.

Create a ⬛ *New Weldment* and 🖫 *Save* it as *welded-assembly-sheet-metal.iam* in the project folder. Place the component *oilpan.ipt* and then a weld seam on the marked position. Repeat the commands for the remaining 3 sheet batches.

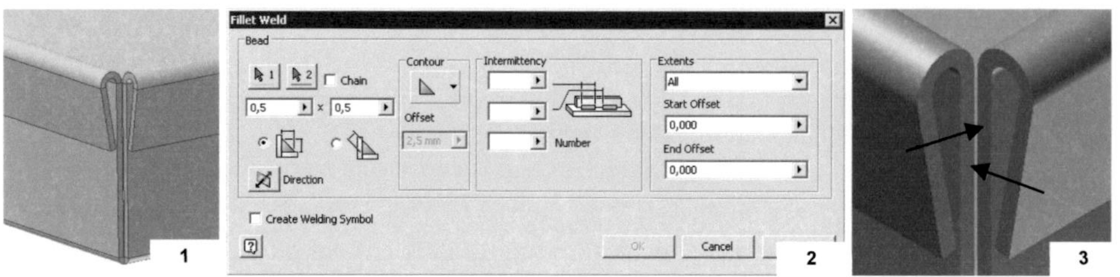

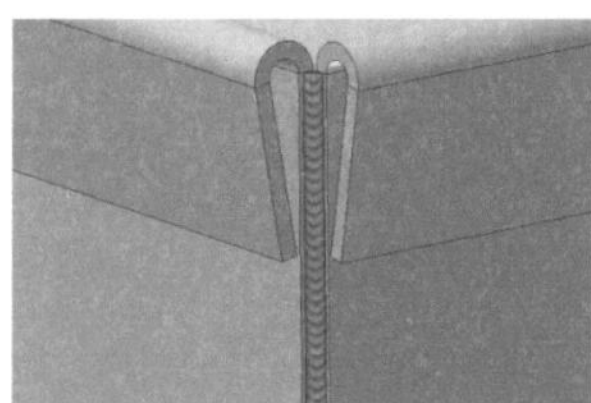

> ⬜ *Welds*
> ◣ *Filled Weld*
> Area 1: Left area (picture 3)
> Area 2: Right area (picture 3)
> Size: 0.5 x 0.5
> ◻ OK

welded-assembly-sheet-metal.iam now is complete. 🖫 *Save* and ✕ *Close* the file.

13 Exercise STRESS ANALYSIS

13.1 Stressanalysis of the crankshaft

The stress analysis of parts is used to test a construction with defined material properties under stress and constraints.

For demonstration purposes we will place a strain on the crankshaft and then analyze the results. Open the part *crankshaft.ipt* and select following commands:

> *Environments* > ▦ *Stress Analysis* > ⬜ *Create Simulation*

Accept the preset data and designate the part a ◈ *Material Assign*.

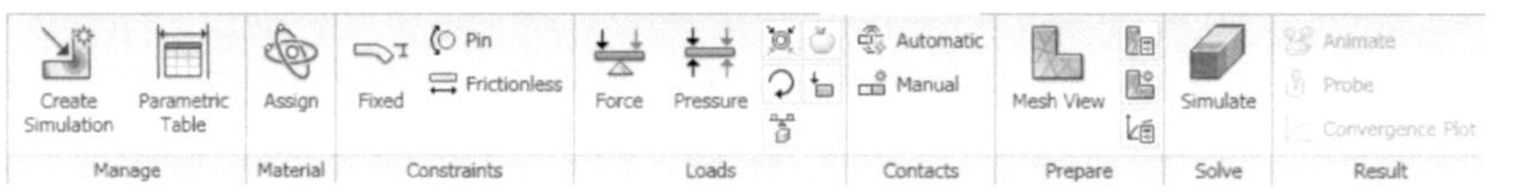

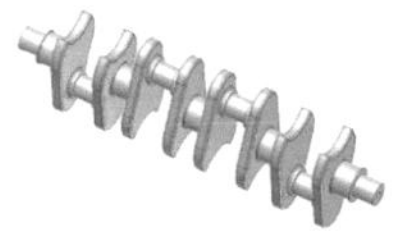

>   **Stress Analysis** (Tab: Environments)
>   **Create Simulation**
> Accept prompts
>   OK

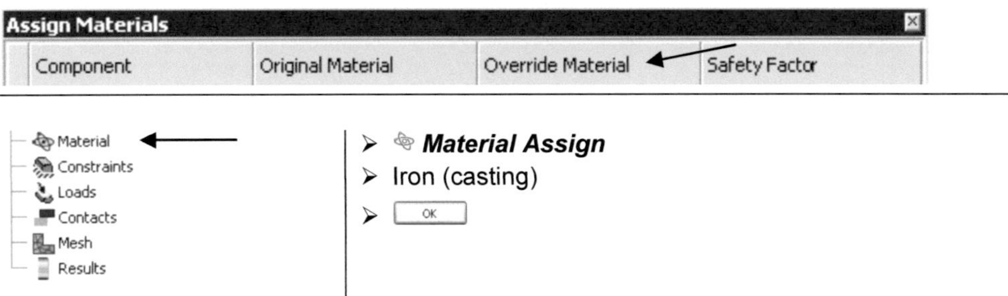

>   **Material Assign**
> Iron (casting)
>   OK

Then the part has to be fixed securely. With the command **Fixed** you clamp a face of the shaft and via the command **Moment** you set a defined torque. **Simulate** then starts the calculations.

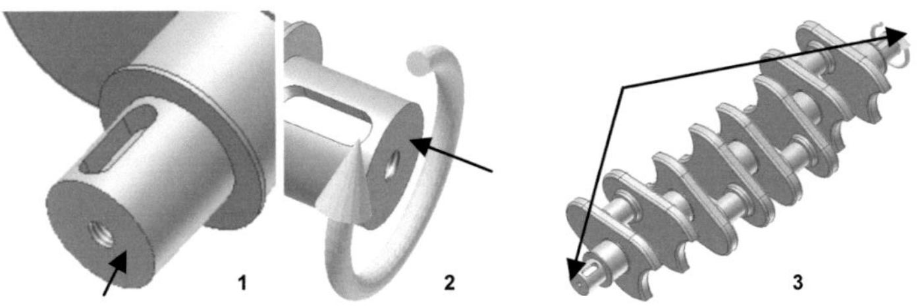

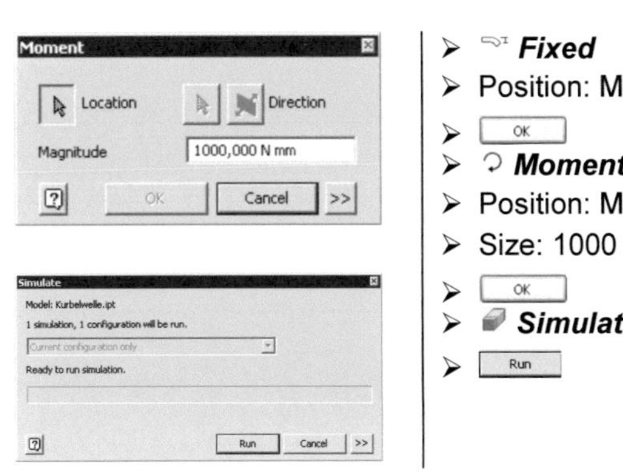

>   **Fixed**
> Position: Marked area picture 1
>   OK
>   **Moment**
> Position: Marked area picture 2
> Size: 1000 Nmm
>   OK
>   **Simulate**
>   Run

To issue all calculations use the command **Report** and type in the required data.

Stress Analysis Report

Analyzed File:	Kurbelwelle.ipt
Autodesk Inventor Version:	2010 (Build 140223002, 223)
Creation Date:	09.11.2010, 14:21
Simulation Author:	Christian
Summary:	

➢   *Report*
➢ General: Type in Title, Author, Name and Path
➢ Properties: Desired issue data
➢ Simulations: Desired simulations
➢   OK

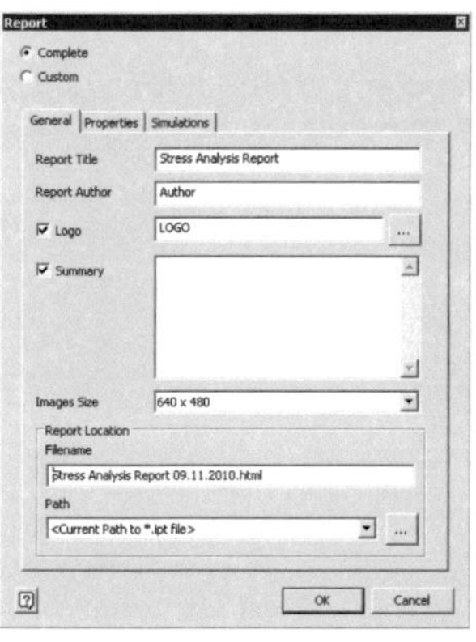

A file with the results of the stress analysis is created.

To enable a simulation of the collected data and receive an optical evaluation of the strains use the command *Animate*.

In the area *Prepare* (*Mesh View*, *Mesh Settings*, *Local Mesh Control*, *Convergence Settings*) you can determine the meshwork size and the fineness of the hubs.

➢   *Animate*
➢   *Start* the simulation in the current window
➢   OK

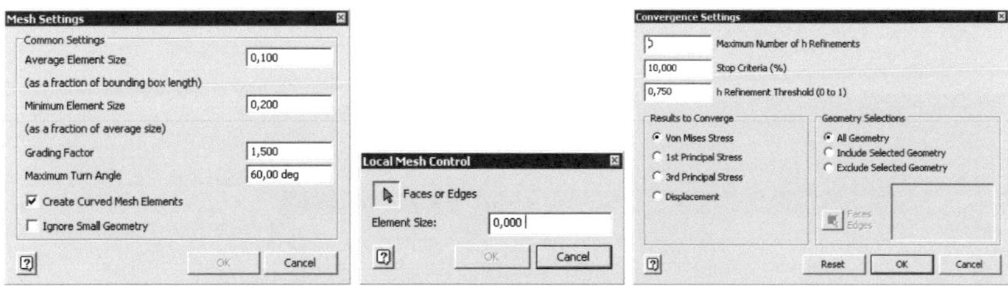

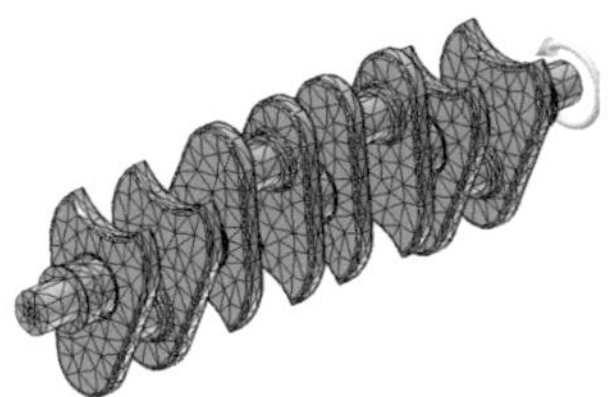

> **Mesh View**
> Creates a view of the FEM-meshwork
> **Mesh Settings**
> Modify the parameters meshwork and fineness degree
> **Local Mesh Control**
> Edit the chosen meshwork settings
> **Convergence Settings**
> Edit the convergence criteria

Tip: With the command ⊛ **Video** you can save the simulation as video file.

The stress analysis now is complete. ⊟ **Save** and ⊠ **Close** the file.

14 Exercise for the area PARAMETERS

Parameters are influential factors which act upon the outside of objects (in our case geometrical outlines). With parameters you can direct the form, size and position of components.

Change values into equations in the area f_x **Parameters** and set geometrical constraints by this. We will create a parametrical sketch and derive resulting components from it. These we then will control via a controlled value.

Create a ⌂ **New Part** and ⊟ **Save** it as **exercise-parameters-ground-sketch.ipt** in your project folder. Then draw two circles (D1= 30, D2= 1, distance = 50).

Open the command f_x **Parameters** and change the values in the area **Parameter Name**.

14.1 Creating the parametrical sketch

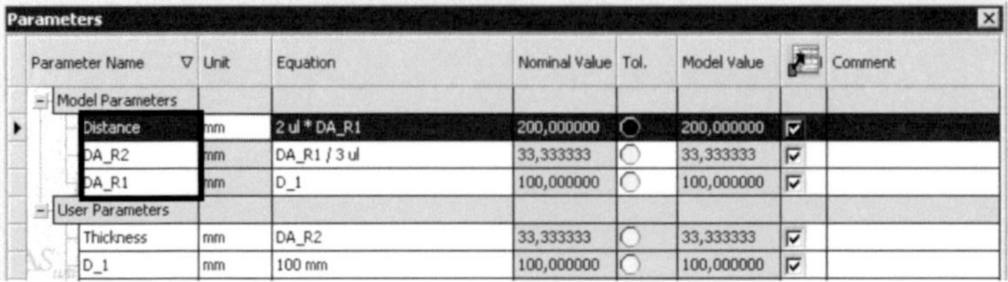

Parameter Name	Unit	Equation	Nominal Value	Tol.	Model Value		Comment
− Model Parameters							
Distance	mm	2 ul * DA_R1	200,000000	○	200,000000	✓	
DA_R2	mm	DA_R1 / 3 ul	33,333333	○	33,333333	✓	
DA_R1	mm	D_1	100,000000	○	100,000000	✓	
− User Parameters							
Thickness	mm	DA_R2	33,333333	○	33,333333	✓	
D_1	mm	100 mm	100,000000	○	100,000000	✓	

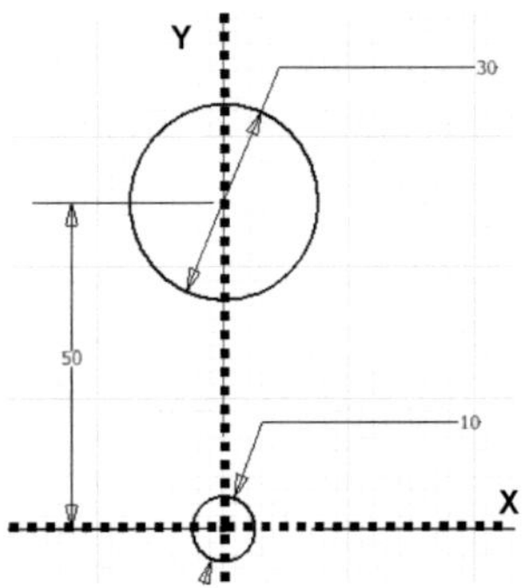

> ➤ **Project Geometry** (X-Y-Z-axes)
> ➤ Draw shown geometry
> ➤ ✔ **Finish Sketch**
>
> ➤ f_x **Parameters**
> ➤ Change following parameter names
> ➤ Parameter with nominal value 50: Distance
> ➤ Parameter with nominal value 30: DA_R1
> ➤ Parameter with nominal value 10: DA_R2

Then insert the two user parameters **D_1** and **Thickness** and replace the values in the area **Equation** by the stated formulas.

Tip: The parameter **D_1** now directs the parameter **DA_R1** (distance between the two centers of the circles). Based on these constraints all other measures are influenced by changes of **D_1**. The sketch now is parametrically dependent.

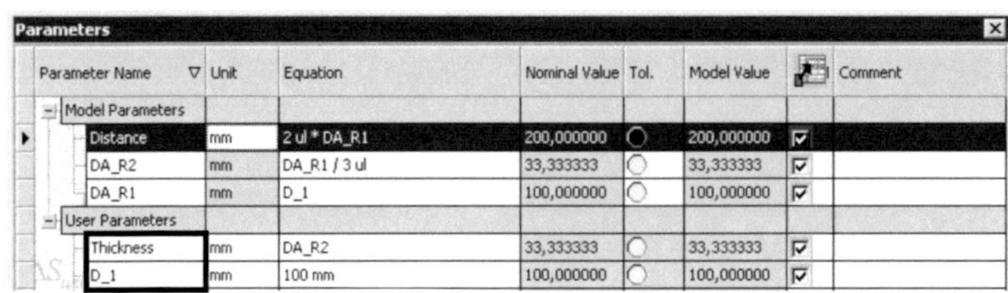

Parameter Name	Unit	Equation	Nominal Value	Tol.	Model Value		Comment
− Model Parameters							
Distance	mm	2 ul * DA_R1	200,000000	○	200,000000	✓	
DA_R2	mm	DA_R1 / 3 ul	33,333333	○	33,333333	✓	
DA_R1	mm	D_1	100,000000	○	100,000000	✓	
− User Parameters							
Thickness	mm	DA_R2	33,333333	○	33,333333	✓	
D_1	mm	100 mm	100,000000	○	100,000000	✓	

> ➤ **Add**
> ➤ User parameter: D_1
> ➤ User parameter: Thickness

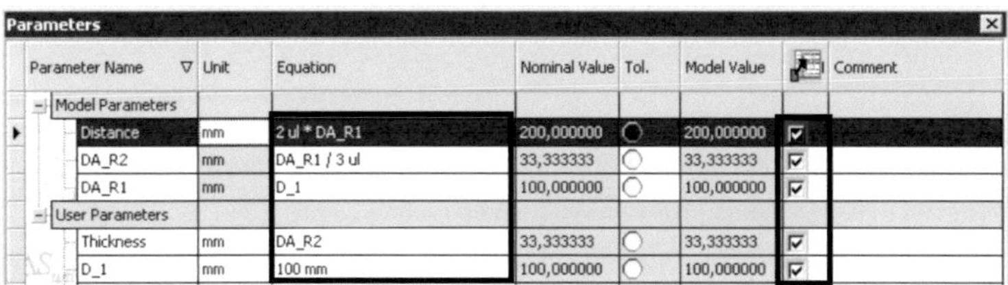

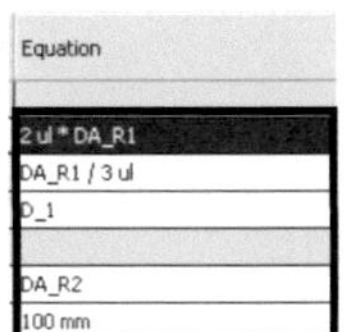

> *Adjust the following equations*
> Parameter Distance: 2 ul * DA_R1
> Parameter DA_R1: D_1
> Parameter DA_R2: DA_R1 / 3 ul
> Parameter D_1: 50
> Parameter Thickness: DA_R2
> OK

Then all parameters have to be marked as **Export Parameter** (set hook on right).

Tip: Consider the order of the entries. You can define any user parameters, however, when you refer to a not (yet) existing parameter in the area equations the value is marked in <u>RED</u> and the system protests.

All parameters used in the area **Equation** have to be already defined. It is better to first define all user parameters and then place the respective equations.

Save and **Close** the file **exercise-parameters-ground-sketch.ipt** now.

14.2 Deriving the pulleys

Create a **New Part** and **Save** it as **exercise-parameters-pulleys.ipt** in your project folder.

Use the command **Derive** to derive the two marked tires from the sketch of the previously created part **exercise-parameters-ground-sketch.ipt**.

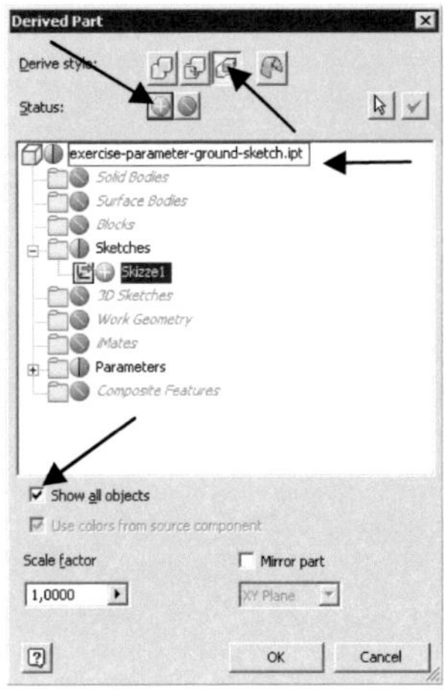

> ✔️ *Finish Sketch*

> 🗐 *Derive* (exercise-parameters-ground-sketch.ipt)
> Mark *exercise-parameters-ground-sketch.ipt* in the new opened window with *LMC*
> 🔘 Add
> Receive 🗐 as volume shape
> Hook at: Show all objects
> Scale factor: 1
> OK

The derived sketch now appears in the drawing area. Besides the sketch also the parameters of the file *exercise-parameters-ground-sketch.ipt* have been assumed and can be used here.

Extrude both circles. The extrusion height in this case is not preset but is directed by the parameter *Thickness*.

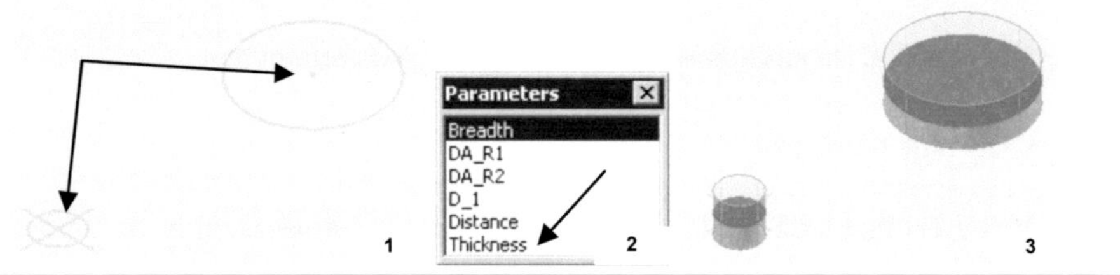

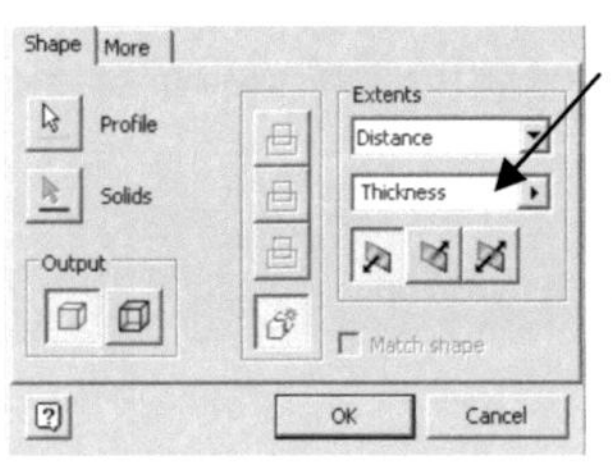

> 🗐 *Extrude*
> Profile: Arc area
> Operation: Join
> Extents: Distance (*Parameters* > *Thickness*)
> Direction: Symmetrically
> OK

💾 *Save* and 💾 *Close* file *exercise-parameters-pulleys.ipt*.

14.3 Creating the parametrical assembly

Create a ⬛ **New Assembly** and save it as **exercise-assembly-parameters.iam** in your project folder.

Place component **exercise-parameters-pulleys.ipt** symmetrically to Y-axis in it (Y-Z-plane of the pulleys should to be placed on the Y-Z-plane of the assembly) and create a new part. Project the marked edges of the small wheel and draw the shown outline.

Finish the sketch and create the shown path for our part in a new ✑ **2D Sketch**.

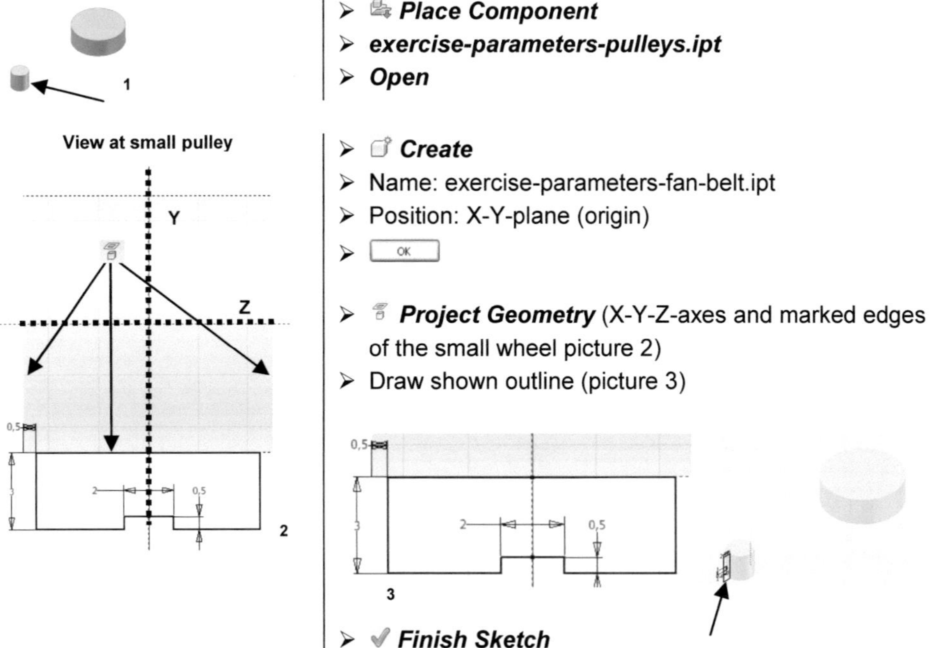

> 🖴 **Place Component**
> exercise-parameters-pulleys.ipt
> **Open**

> ✒ **Create**
> Name: exercise-parameters-fan-belt.ipt
> Position: X-Y-plane (origin)
> OK

> ☷ **Project Geometry** (X-Y-Z-axes and marked edges of the small wheel picture 2)
> Draw shown outline (picture 3)

> ✔ **Finish Sketch**

Tip: The sketch is located to the small pulley on the bottom side. Make sure that the outline with the ∟ **Constraint Coincidence** is located towards the Y-axis, the entire outline is symmetrical and the upper line with the ⤫ **Constraint Collinear** is attached to the bottom outside edge of the small wheel.

To clarify the parametrical change of the belt in constraint to the size of the pulleys later on, fixed values for belt height (3) and the dimensions of the notch (2 x 0.5) have been used. This ways the proportions during the changes become more visible.

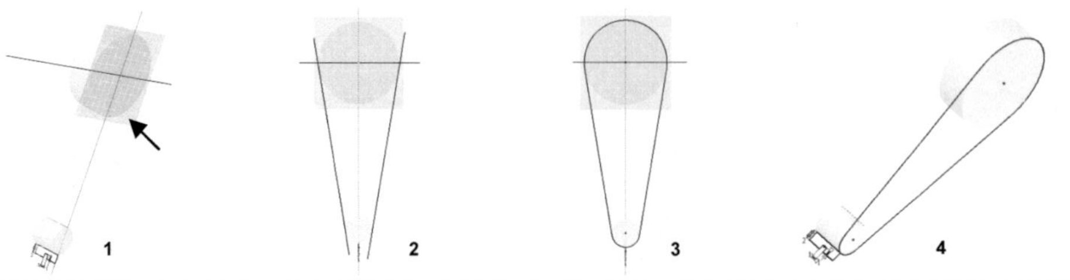

> ⬚ **2D Sketch** (marked area picture 1)
> ⬚ **Project Geometry** (X-Y-Z-axes and both circles)
> Mark projected lines as ⟍ **Construction Lines**
> Draw 2 tangents and two arcs (center point), like chapter 7.6.12
> ✔ **Finish Sketch**

Use the command ⬚ **Sweep** to generate the volume shape. Then leave the editing area of the new part and go back to the superior assembly.

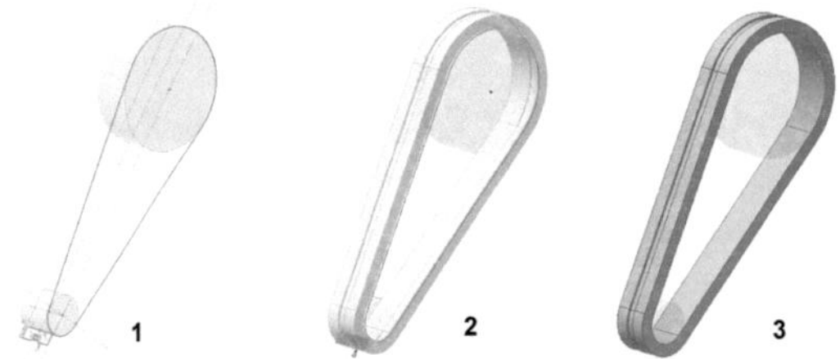

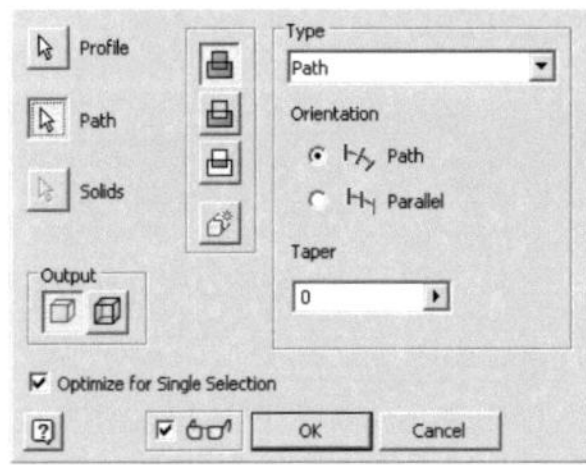

> ⬚ **Sweep**
> Profile: 1. Created sketch
> Path: 2. Created sketch
> Type: Path
> Operation: Join
> [OK]

Now ⬚ **Step Higher**, ⬚ **Save** the file **exercise-assembly-parameters.iam** and all subordinate files, but do **NOT** close.

Open the file **exercise-parameters-ground-sketch.ipt** and in the area f_x **Parameters** change the user parameter **D_1** to 5.

Then switch to *exercise-assembly-parameters.iam* and use the command 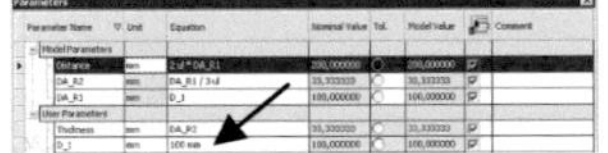 *Local Update*. The changes in the sketch resulted in a change of the wheels and of the belt in the assembly.

Tip: Should there be no changes in the assembly after changing the sketch, save the sketch file according to value changes and then switch to the assembly.

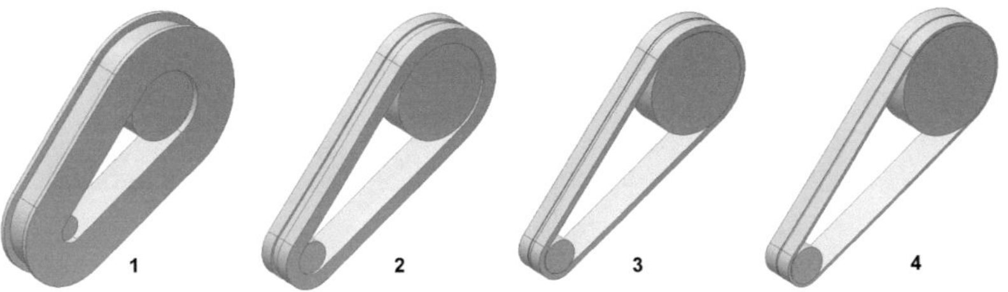

> f_x *Parameters*
> User parameter D_1: 5
> [OK]

Switch to the file *exercise-assembly-parameters.iam* (tab is located in the left bottom portion of the drawing window)

> *Local Update*

Assembly will now refresh the changes in the file *exercise- assembly-parameters.iam* and show new. Repeat these changes in the parameters *D_1* in the file *exercise-parameters-ground-sketch.ipt* with the following values:

> User parameter D_1: 20
> User parameter D_1: 50
> User parameter D_1: 100

14.4 Directing the values via table conjunction

Then we join the file *exercise-parameters-ground-sketch.ipt* with an excel table. This way we can change the parametrical values without having to edit the actual file.

For this we create an excel-file in our project folder and 🖫 *Save* this as *exercise-parameters.xls*.

➢ Open the project folder (desktop)
➢ **RMC > New > EXCEL-worksheet**
➢ 💾 **Save** as **exercise-parameters.xls**
➢ 📂 **Open** the newly created table

Type in following values:

➢ Cell A1: EXCEL_DA_R1
➢ Cell A2: 100
➢ 💾 **Save**

Open the file **exercise-parameters-ground-sketch.ipt** and join the file with the table.

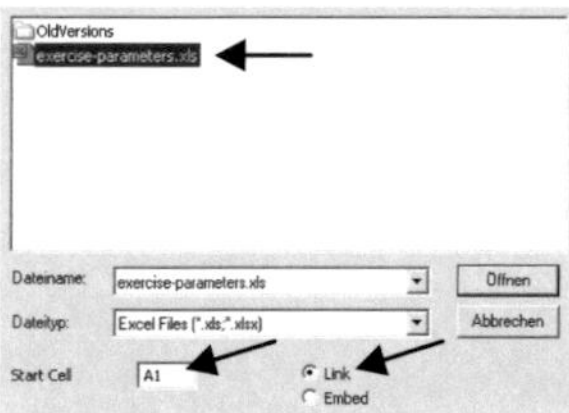

➢ f_x **Parameters**
➢ Link
➢ **Exercise-parameters.xls** (in the project folder)
➢ Hook at: Link
➢ Start line: A1
➢ **Open**

Under user parameter **D_1** now enter following equation:

➢ Equation: EXCEL_DA_R1

Done

💾 **Save** and ✖ **Close** the file **exercise-parameters-ground-sketch.ipt**. Reopen the table **exercise-parameters.xls** and change the value in cell **A2** to **5**.

💾 **Save** the table and open the file **exercise-assembly-parameters.iam**.

When you chose the command 🖳 **Local Update** you can see that the components now are influenceable via the changes in the table and a change of the values in the file **exercise-parameters-ground-sketch.ipt** is no longer necessary.

Tip: Besides MS-Excel of course all other table programs (e.g. Open Office) can be used. It is just important to immediately save changes to the table.

15 PACK and GO - Archiving and Copying

In our last exercise we will cover the proper archiving of projects. Extensive projects can be completely copied, archived or passed on with the command ▩ *Pack and Go*. Not only parts and assemblies are copied here, but also their references, special styles and colors and plenty more. It is important that all references, joining, structures and other needed files are included.

First create a folder *Export* in your project folder. ▣ *Open* the file *assembly-four-stroke-engine.iam* and choose the commands:

> ➢ ▣ *IPro* > ▣ *Save as* > ▩ *Pack and Go*

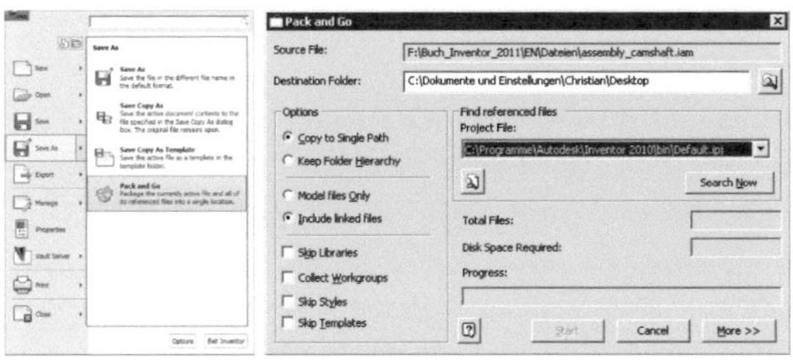

> ➢ ▣ *IPro*
> ➢ ▣ *Safe As*
> ➢ ▩ *Pack and Go*
>
> ➢ Source file: assembly-four-stroke-engine.iam
> ➢ Target folder: Folder *Export* in project folder
> ➢ Options: Keep folder structure
> ➢ Include linked files
> ➢ Project File: four-stroke-engine.ipj (in project folder)
> ➢ Start

In the chosen folder *Export* all necessary files are contained and in the text-file *packngo.txt* you find an overview.

You now can copy the entire folder and open the entire project including parts, assemblies, drawings, links, styles, colors and other project-relevant data on any computer with *Autodesk® Inventor® 2011* installed and use.

Thank you.